Advanced Thermal Management Materials

T0140452

Guosheng Jiang • Liyong Diao • Ken Kuang

Advanced Thermal Management Materials

 Springer

Guosheng Jiang
College of Materials Science and
  Engineering
Central South University
Changsha, Hunan, China

Liyong Diao
Brewer Science
Springfield, MO, USA

Ken Kuang
Torrey Hills Technologies
San Diego, CA, USA

ISBN 978-1-4899-9254-3   ISBN 978-1-4614-1963-1 (eBook)
DOI 10.1007/978-1-4614-1963-1
Springer New York Heidelberg Dordrecht London

Printed on acid-free paper

Springer is part of Springer Science+Business Media (www.springer.com)

Preface

Microelectronics packaging was born out of necessity when the integrated circuit (IC) was invented in 1947. Microelectronics packaging is interdisciplinary in nature and involves physics, chemistry, materials science, mechanical engineering, electrical engineering, and more. Microelectronics packaging refers to the enclosure of electronic devices or ICs according to the requirements of each component to achieve a reasonable layout, assembly, bonding, and connection. It is important since it protects devices from moisture, heat, mechanical stresses, mechanical shock, thermal shock, and chemical erosion/corrosion.

Nowadays the trends in electronics are toward a small footprint, light weight, low cost, high performance, and high reliability. All of these trends have led to highly integrated ICs and sophisticated packaging schemes. When the chip power density increases, it is critically important to dissipate the extra heat efficiently. If the heat is not managed properly, the chip's working environment is worsened, leading to higher chip operating temperatures and unstable device performance. In extreme cases, the chips burn out, resulting in fire and safety hazards. Statistically speaking for semiconductors, for every 18 °C operating temperature increase, chip reliability is reduced by two to three times. Therefore, it is very important to manage the waste heat properly.

Different thermal management materials are used in microelectronics packaging. They typically possess a high thermal conductivity (TC) and a low coefficient of thermal expansion (CTE) and are used mainly to dissipate heat and to provide structural support. Traditionally they are also called heat sink materials.

There are many ways to manage waste heat, such as cryogenic coolers, active chilled water pipes, and cooling fans. Most of these methods focus on external heat management, i.e., on how to dissipate waste heat from the environment. There is another fundamental challenge, which is how to dissipate the heat from the IC active layer itself. This is typically done by conduction using heat sink materials.

Lately there has been a great deal of active research on thermal management materials, especially on dielectrics, metals, and metal matrix composites. Results of current research are typically dispersed in various technical journals and conference proceedings. Thermal management engineers have been hard pressed to find a comprehensive and practical book to cover both the fundamentals of thermal management and selection guidelines, an issue this book hopes to address.

The main objective of this book is to introduce various thermal management materials and their fabrication methods. Most of the materials covered are based on our 10+ years of direct R&D and manufacturing experience. We hope to provide an effective reference book for thermal management engineers and packaging engineers.

The book is divided into ten chapters. Chapters 1, 2, and 3 cover the basics of thermal management and traditional thermal management materials. Chapters 4, 5, 6, 7, and 8 cover copper- and aluminum-based thermal management materials. Finally, Chaps. 9 and 10 discuss the application of these materials in laser diodes and future development trends.

It is our goal to introduce the reader to thermal management basics, theory, and application. At the same time, we strive to reference as many of the latest research papers as possible so that readers can attain a comprehensive understanding of the current status and future trends of the field. As stated previously, thermal management materials are being actively studied worldwide. We cannot possibly cover all the latest developments and welcome feedback from readers like you.

During the writing of this book, we received generous support from Prof. Renzheng Tang, Prof. Zhifa Wang, and Dr. Yi Gu and from graduate students Hong Wu, Dexin Chen, Jun Zhou, Qiwang Zhang, and Xing Yu from Central South University. Without their support and guidance, this book could not have been written. In addition, we would like to extend our appreciation to Ganesh Hariharan and Michael Shaw from Torrey Hills Technologies, LLC, and Adam Ding from the University of Southern California. They rendered valuable assistance in providing materials for several chapters.

Changsha, Hunan, China   Guosheng Jiang
Springfield, MO, USA   Liyong "Alex" Diao
San Diego, CA, USA   Ken Kuang

Contents

Chapter 1
Introduction to Thermal Management in Microelectronics Packaging

Abstract Heat generated by electronic devices and circuitry must be dissipated to improve reliability and prevent premature failure. Thermal management goes hand in hand with microelectronics packaging. In this chapter, we will present the motivations and the basic concepts of thermal management by heat sink materials, such as heat flux, thermal resistance, and thermal circuits. Next we will introduce the levels and classifications of packaging and the functions of microelectronics packaging. Finally, we will introduce the development stages of thermal management materials.

1.1 Basics of Thermal Management in Microelectronics Packaging

In extreme environments, temperatures can reach up to 1,000°C and higher. The power density in some computer servers can reach up to 5 W/cm^2. The junction temperature is the highest temperature of the semiconductor in an electronic device. It is generally accepted that the maximum junction temperature is about 150°C for silicon devices. It is desirable to keep it below 85°C. The maximum junction temperatures allowed for SiC and GaN materials are approximately 750 and 350°C, respectively. Statistically, the reliability of integrated circuit (IC) chips is reduced by two to three times for every 18°C increase in chip temperature.

There are many ways to reduce the IC chip operating temperature, e.g., cryogenic coolers, active chilled water pipes, cooling fans, and heat sinks. The cooling of optics and detectors to reduce signal noise is one of the important applications of cryogenic cooling technologies both today and for the near future.

One of the most important functions of thermal management materials in microelectronic packaging is the efficient transfer of heat from the semiconductor junction to the ambient environment. This process can be separated into three phases:

- Heat transfer within the semiconductor component package;
- Heat transfer from the package to a heat dissipater;
- Heat transfer from the heat dissipater to the ambient environment.

G. Jiang et al., *Advanced Thermal Management Materials*,
DOI 10.1007/978-1-4614-1963-1_1, © Springer Science+Business Media New York 2013

1.1.1   Heat Flux

Radiation, conduction, and convection are three ways to dissipate heat from a device. The rate at which heat is conducted through a material is proportional to the area normal to the heat flow and to the temperature gradient along the heat flow path. For a 1D, steady-state heat flow the rate is expressed by Fourier's equation:

$$\frac{Q}{A} = k\frac{\Delta T}{d}$$

where:

k=thermal conductivity, W/m-K,
Q=rate of heat flow, W,
A=contact area,
d=distance of heat flow,
ΔT=temperature difference.

1.1.2   Thermal Resistance

Thermal resistance is the measure of a substance's ability to dissipate heat, or the efficiency of heat transfer across the boundary between different media. A heat sink with a large surface area and good air circulation (airflow) gives the best heat dissipation:

$$R = A\frac{\Delta T}{Q} = \frac{d}{k}$$

For homogeneous materials, thermal resistance is directly proportional to thickness. For nonhomogeneous materials, the resistance generally increases with thickness, but the relationship may not be linear.

The thermal impedance, Θ, of a material is defined as the sum of its thermal resistance and any contact resistance between it and the contacting surfaces:

$$\Theta = R_{material} + R_{contact} = R_{material} + (T_a - T_b)/Q,$$

where T_a and T_b are temperatures at the interfacing boundaries.

Surface flatness, surface roughness, clamping pressure, material thickness, and compressive modulus have a significant effect on contact resistance. Because these surface conditions can vary from application to application, the thermal impedance of a material will also be application dependent.

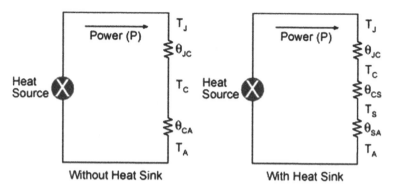

Fig. 1.1 Thermal circuit models [1]

1.1.3 Thermal Management Examples with a Heat Sink and Without a Heat Sink

To simplify the thermal transport concept, we use the thermal circuit model, which is similar to the electric circuit model. The through circuit element is the power, and the potential element is the temperature. The circuits shown in Fig. 1.1 represent two cases (Table 1.1):

- *Case 1*: the bare die is connected to the case and the case is exposed to the environment;
- *Case 2*: the bare die is connected to the case, then to a heat sink, and the case is exposed to the environment.
- Junction temperature without a heat sink:

$$T_J = T_a + Q \times \theta_{JA} = 50 + 20 \times 4.7 = 148°C$$

- Junction temperature with heat sink and its heat resistance equals 1.35°C/W and has an interface material with a heat resistance of 0.13°C/W.

$$T_J = T_a + Q \times (\theta_{JC} + \theta_{CS} + \theta_{SA}) = 50 + 20(0.13 + 0.1 + 1.35) = 81.6°C.$$

From the two simple cases we understand that a well-chosen heat sink material could lower the junction temperature and significantly improve the reliability of the device performance because the junction temperature is controlled under 85°C.

1.2 Heat Generation and the Purpose of Thermal Management in Microelectronics Packaging

Since the inception of semiconductor devices in 1948, the miniaturization of electronic components, microminiaturization, and the integration of technology, especially the emergence of new chip materials (such as SiC and GaN) and their

Table 1.1 Thermal circuit parameters [1]

Parameters	Name	Unit	Description	Example
Θ_{JA}	Θ from junction to ambient	°C/W	Specified in data sheet	4.7°C/W
Θ_{JC}	Θ from junction to case	°C/W	Specified in data sheet	
Θ_{CS}	Θ from case to heat sink	°C/W	Thermal interface material thermal resistance	
Θ_{CA}	Θ from case to ambient	°C/W		
Θ_{SA}	Θ from heat sink to ambient	°C/W	Specified by heat sink manufacturer	0.13°C/W
T_J	Junction temperature	°C	Junction temperature as specified under recommended operating conditions for device	
T_{JMAX}	Maximum junction temperature	°C	Maximum junction temperature as specified under recommended operating conditions for device	85°C
T_A	Ambient temperature	°C	Temperature of local ambient air near component	
T_{Amax}	Maximum ambient temperature	°C	Maximum temperature of local ambient air near component	50°C
T_S	Heat sink temperature	°C		
T_C	Device case temperature	°C		
Q	Power	W	Total power from operating device; use estimated value for selecting a heat sink	20 W

applications, have led to an increase in electronic equipment power density and required greater heat dissipation in high-power FR, microwave, and millimeter-wave devices and on-board and satellite electronic equipment. As the temperature of semiconductor devices increases by 10°C, their reliability is reduced by 50 %. In 2000, the heat flux of large computer chips has exceeded 100 W/cm². At present, it reaches a level of 300 W/cm². The main methods for cooling electronic equipment include natural cooling, forced air/liquid cooling, cold plate, phase-change cooling, heat pipes, and so on.

Electronic packaging refers to the enclosure of electronic devices or ICs according to the requirements of each component to achieve a reasonable layout, assembly, bonding, connecting, and operating environment. It also includes process isolation and protection to prevent the buildup of moisture, dust, and harmful gases on electronic devices or ICs, to prevent external damage, and to stabilize component parameters. Thermal management of electronic packaging concerns the temperature control of heat-consumption electronic devices, unit systems, or whole systems by reasonable cooling, heat-dissipation, and structural-design optimization to ensure normal and reliable operations. Thermal management materials function as the substrate, or base, of electronic components; they cool and support electronic components.

In addition, such materials conduct heat generated by electronic components to the exterior environment to control the temperature of chips so as to improve the chips' life span and reliability. As a result, many thermal management materials are referred to as heat sink materials or cooling materials in the literature. The key to improving thermal cooling performance is the thermal properties of thermal management materials. They play an important role in the cooling of electronic components. Before the details of thermal management materials are discussed, we must introduce basic concepts in electronic packaging technology.

1.3   Introduction to Electronic Packaging

Packages are commonly divided into four levels: zero-level, first-level, second-level, and third-level packages. Zero-level packages include interconnecting and encapsulation of IC chips. First-level packaging refers to packaging of a single-chip module (SCM) into a multichip module (MCM) involving a wire interconnect, tape interconnect, case materials, lid seal, and lead. Second-level packaging consists of packaging of SCMs, MCMs, and connectors into printed circuit boards (PCBs) involving reinforcement fiber materials, resins, laminates, flexible printed board materials, and conformal coating. Finally, third-level package refers to the packaging of PCBs, cables, power supplies, and ancillary systems into a frame or box (e.g., motherboard) using back-panel materials, connector materials, cables, and flex circuit materials.

The requirements for packaging materials differ at every level. At the first level, the packaging materials must have a high density and thermal conductivity, where the coefficient of thermal expansion (CTE) of the materials matches that of the chip or ceramic. At the second level, the packaging materials must have a cooling capacity, a low CTE, high hardness, and high damping. At the third level, the packaging materials must be lightweight, vibration resistant, and heat dissipating. At the zero and first levels, the packaging is completed by the component developer and producer, whereas at the second and third levels, the unit manufacturer fulfills this function.

1.3.1   Classification of Electronic Packaging

There are several classifications for electronic packaging: metal, ceramic, metal ceramic, and plastic packages. Depending on shape, size, and structure, it can be classified into pin packages, surface mount packages, area array packages, advanced packaging (3D packages), and so on. Based on the field of application, it is divided into microwave power device packaging, large-scale integrated circuit packaging, optical packaging, MEMS packaging, and high-temperature packaging.

1.3.2 Functions of Electronic Packaging

Electronic packaging has four main functions: physical protection, power and signal distribution, cooling, and mechanical support. Details follow.

1.3.2.1 Physical Protection

Chips must be isolated from the external world to prevent electrical circuit corrosion by impurities in the air, to minimize functional degradation caused by corrosion, and to protect the very delicate chip surface and connecting leads from electrical or thermal damage by external forces and the external environment.

1.3.2.2 Power and Signal Distributions

Interconnecting conductors deliver power and signals to chips so that chips can fulfill their function. Wiring length should be minimized. For high-frequency signals, propagation delay and cross talk need to be minimized. Chips and their packaging should be properly grounded.

1.3.2.3 Cooling Channel

The cooling channel relates to how the heat generated by electronic devices is transferred to the external heat transfer channel. Based on the requirements for heat dissipation, thinner packaging is better. When the chip power is greater than 2 W, a heat sink or heat spreader is needed to enhance the package's heat-dissipation cooling; when the chip power is greater than $5 \sim 10$ W, forced cooling must be used.

1.3.2.4 Mechanical Support

Packaging for chips and other electronic components provide solid and reliable mechanical support. At the same time, matching the CTE of the packaging with that of the frame or substrate can ease stress resulting from external environmental change and thermal stress from the chips, thereby preventing chip damage and failure.

1.4 Development Trends of Electronic Packaging Technology

Electronic packaging technology continues to progress, propelled by miniaturization, high performance, versatility, and the low-cost requirements of electronic devices. The development trends are as follows:

- Electronic packaging technology continues to move in the direction of ultrahigh density. 3D packaging, MCM packaging, system in package (SIP), and other high-density packages are developing rapidly.

- The miniaturization of electronic packages is progressing rapidly. Chip-size ultrasmall packaging — wafer-level packaging — is evolving.
- Electronic packaging technology continues to develop from SCMs to MCMs. In addition to MCMs, there are multichip packages (MCPs), SIP, and stacked packaging.
- Electronic packaging technology is progressing from separate units to systems such as system on chip (SOP), SIP, and other packages.
- Electronic packaging technology continues to develop in the direction of high performance, multiple directions, high frequency, and high power.
- Electronic packaging technology is evolving in the direction of high integration, including board-level integration and chip-level integration.

1.5   Development Stages of Thermal Management Materials

With the development of electronic packaging technologies, the power of components has increased, and there is an ever-increasing number of electronic packaging materials. Generally speaking, the development of thermal management materials is divided into the following three stages.

1.5.1   Low-power-device Thermal Management Materials

Low-power-device packaging materials, also known as the first generation of electronic packaging materials, mainly satisfy the packaging needs of low-power devices. There are two classes. The CTE of the first class matches that of Si glass. Because the devices have low power, the requirement for thermal conductivity and matching CTE is less strict. The most widely used materials are Kovar, Invar alloys, and epoxy resin packages (Fig. 1.2).

   Another class of low-power-device packaging material is pure metals such as Fe, Cu, W, Mo, and others. Copper has good thermal conductivity and is inexpensive, which is why it is widely used. Many transistor outline headers still use copper as a base plate (Fig. 1.3). Semiconductor devices in power electronics are still widely used in pure Mo films as a Si-chip support base plate (Fig. 1.4).

1.5.2   High-power-device Thermal Management Materials

Since the 1990s, as microwave, radio-frequency electronic packaging technologies have developed, the density of chip integration has increased dramatically. Transistor gate lengths are becoming increasingly shorter and are now at 0.026 µm. The output power of silicon devices is also increasing. With the use of GaAs chip materials,

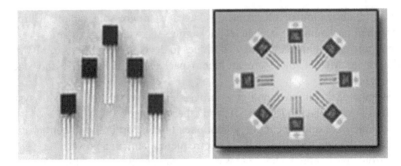

Fig. 1.2 Transistor components encapsulated with epoxy resin

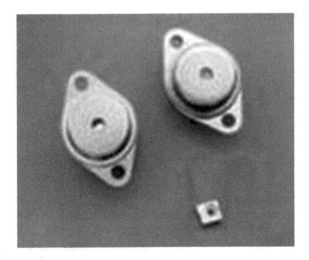

Fig. 1.3 TO header tube socket made of copper material

Fig. 1.4 Silicon chip rectifier SCR and the support base made of pure Mo material

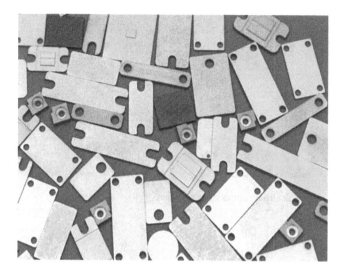

Fig. 1.5 Second-generation thermal management materials

packaging technology has moved away from minimizing surface mount technology in the direction of 3D microelectronic packaging technology. The increase in chip power density requires that the accompanying thermal management materials have higher thermal performance. Such materials are called the second generation of thermal management materials. They include, e.g., W/Cu, Mo/Cu, AlSiC, and AlSi (Fig. 1.5). Second-generation thermal management materials have a higher thermal conductivity, and their CTE matches well that of the contacting materials. Most of these materials are composites; single-phase metals cannot satisfy the requirements of high-power chips.

1.5.3   Ultra-high-power-device Thermal Management Materials

With the advent of high-power-chip SiC materials, thermal management materials have demanded higher thermal conductivity. The second-generation thermal management materials cannot satisfy the thermal conductivity requirements of SiC chips. As a result, third-generation thermal management materials have been developed. They include composite materials made with diamond and C fiber with very high thermal conductivity. The commercialized representatives include diamond/copper, diamond/aluminum, carbon fiber copper, and carbon fiber aluminum (Fig. 1.6).

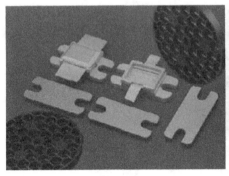

Fig. 1.6 High-power electronic component packages

References

1. http://www.altera.com/support/devices/power/thermal/pow-thermal.html
2. White M, Cooper M, Chen Y (2003) Impact of junction temperature on microelectronic device reliability and considerations for space applications. Integrated reliability workshop final report, 2003 IEEE International
3. Wrzecionko B, Biela J, Kolar JW (2009) SiC power semiconductors in HEVs: influence of junction temperature on power density, chip utilization and efficiency. In: Proceedings of the IECON 2009, Porto, Portugal, 3–6 Nov 2009
4. Thermal Considerations for GaN Technology, www.nitronex.com/pdfs/AN-012%20 Thermal.pdf
5. Pecht M et al (1999) Electronic packaging materials and their properties. CRC Press, Boca Raton
6. Brooks C, Choudhury A (2002) Failure analysis of engineering materials. McGraw-Hill, New York
7. Weiming Z, Hongyu Z, Wei Y (2010) Electronic packaging and micro-assembly of characteristics and development trend of the sealed. Defense Manufact Technol 2(1):60–62
8. Avram BC, Lyengar M, Kraus AD (2007) Design of optimum plate fin naturalconvection heat sinks. ASME Trans J Electron Packag 125(2):208–216
9. Sikka KK (1997) Heat spreaders and heat sinks for mixed convection electronic cooling. UMI Company, Michigan
10. Jacobs HR, Hartnett JP (1991) Thermal engineering: emerging technologies and critical phenomena. Workshop Report, NSF Grant No. CTS291204006, 1392176
11. Shou-guang Y, Ma Zhe (2003) tree, rollin, Chen ice. Electrical and electronic equipment efficient heat pipe cooling technology research and development. Jiangsu University (Natural Science) 8(17): 9212
12. Dage H, Shuanggeng Y (2006) Cooling Techniques for High Flux Electronics. Fluid Mach 34(9):71–74
13. Varsamopoulos G, Abbasi Z, Gupta SKS (2010) Trends and effects of energy proportionality on server provisioning in data centers. In: Proceedings of the 2010 international conference on high performance computing (HiPC), Tempe, AZ, USA, 19–22 Dec 2010
14. ASHRAE (2005) Datacom equipment power trends and cooling applications. Atlanta, GA

Chapter 2
Requirements of Thermal Management Materials

Abstract In this chapter, we will present the requirements of thermal management materials from a physics point of view. First, the mechanism of a metal electron and the mechanism of a metal lattice on thermal conductivity are discussed in detail. Next, the effects of atomic structure, chemical composition, porosity, and temperature on thermal conductivity are presented. In the following section, we will introduce methods to measure thermal conductivity, the coefficient of thermal expansion, and hermeticity. Finally, the emergence of quality requirements for thermal management materials are discussed.

To fulfill the functions of mechanical support, electric connection, heat transfer, and protection of microchips, thermal management materials must meet the requirements described in the following sections.

2.1 Thermal Conductivity Requirements of Thermal Management Materials

The main purpose of thermal management materials in electronic components is the timely transfer of heat generated by microchips through external heat-spreading channels. The pros and cons of thermal management materials play a very important role in the life span and reliability of electronic components. Metallic materials like copper, silver, and aluminum have good thermal conductivity. Copper has a thermal conductivity of up to 400 W/mK. Good thermal conducting oxides and ceramic materials include beryllium oxide, silicon carbide, and aluminum nitride. Aluminum oxide is one of the most widely used materials; unfortunately, its thermal conductivity is not good enough. 96% Al_2O_3 has a thermal conductivity of 20 W/mK at 25°C. The thermal properties of common thermal management materials are listed in Table 2.1.

G. Jiang et al., *Advanced Thermal Management Materials*,
DOI 10.1007/978-1-4614-1963-1_2, © Springer Science+Business Media New York 2013

Table 2.1 Thermal conductivity of thermal management materials

Material	Thermal conductivity (W/mK)	Material	Thermal conductivity (W/mK)
Si	150	W	174
Ge	77	Mo	140
SiC	270	Invar	11
GaAs	45	Kovar	17
SiGe	150	Al	230
InP	97	Cu	400
GaP	133	Au96.85%-Si	27
GaN	13–33	Au80%-Sn	57
InAs	35	Au88%-Ge	44
InSb	19	W85-Cu	200
96%Al_2O_3	20	Mo85-Cu	170
AlN	270	CMC1:1:1	280
BeO	210	CPC1:1:1	300
Epoxy	1.7	SiC70%-Al	160
CVD Diamond	1,300–2,000	Si75%-Al	150
C_f/Al	350	C_f/Cu	415
Diamond/Al	500	Diamond/Cu	600

2.1.1 Heat-Conducting Mechanism

When two pieces of metal or two objects with different temperatures contact each other, heat energy shall transfer from the high-temperature piece to the low-temperature piece. This is the thermal conducting phenomenon. The process of heat conduction is a process of energy transport. In a solid, energy may be carried by free electrons, lattice vibration waves (phonons), and electromagnetic radiation (photons). Therefore, solid-state thermal conducting can be accomplished by electrons, phonons, and photons. For insulators, the phonon is the main carrier; for pure metals, the electron is the main carrier; for alloys, lattice wave plays some role in addition to electrons. As a result, for the materials discussed previously, metals have the highest thermal conductivity, alloys are next, and semiconductors have the lowest thermal conductivity.

Another commonly used engineering parameter related to thermal conductivity is thermal diffusivity. It is defined as

$$\alpha = \frac{\lambda}{dc}, \tag{2.1}$$

where α is the thermal diffusivity, d the density, c the specific heat capacity, and λ the thermal conductivity.

The physical meaning of thermal diffusivity is linked to a transient heat conduction process. Transient heat conduction is a process involving both temperature change and heat transfer. Thermal diffusivity is the physical link between the two. It marks the rate of temperature change. Under identical heating and cooling conditions, the greater the thermal diffusivity, the smaller the temperature difference throughout the object. For example, in a quenching process, the outside steel temperature is outside, while the internal temperature is high. If the thermal conductivity is high, the temperature gradient will be small and the sample temperature uniform; on the other hand, the temperature difference between samples will be large. Because the density and the specific heat capacity of steels are quite similar, it is generally accepted that the higher the thermal diffusivity, the higher the thermal conductivity.

Thermal conductivity is an engineering parameter used to judge the utility of insulation or heat exchange materials. It is also an important parameter for calculating the holding time of material heat treatments. Thermal conducting is a complex problem that can be affected by many factors. Generally speaking, it is a nonequilibrium problem. Here we make only a preliminary introduction.

2.1.1.1   Mechanism of Metal Electron Thermal Conductivity

According to the electron theory of metals, a large number of free electrons in metals can be regarded as free-electron gas. It is a reasonable approximation to borrow the ideal gas equation to describe the thermal conductivity of the thermal conductivity of free electrons. The expression for ideal gas thermal conductivity is

$$k = \frac{1}{3} cvl \tag{2.2}$$

where c is the gas heat capacity per unit volume, l the mean free path of the particle, and v the average particle velocity.

Taking into account various mechanisms, the thermal conductivity of the solid λ is expressed as

$$\lambda = \frac{1}{3} \sum_{j}^{n} c_j v_j l_j, \tag{2.3}$$

where the subscript j represents the type of heat carrier, c_j and l_j represent various heat carriers' heat capacity per unit volume and the mean free path, and v_j is the carrier fluid velocity. If the heat carrier is a lattice wave, then v_j is the corresponding group velocity of the lattice wave.

A thermal conducting carrier in a metal is mainly free electrons; the lattice wave also plays a role. As a result, total thermal conductivity can be written as

$$\lambda = \lambda_e + \lambda_a \tag{2.4}$$

where λ_e is the electron thermal conductivity and λ_a is the lattice thermal conductivity. For pure metals, the thermal conductivity is determined mainly by free electrons; for alloys, the phonon thermal conductivity contribution needs to be considered.

For electronic thermal conductivity, the electron mean free path L is determined entirely by scattering of electrons in metals. If the metal lattice is complete, then

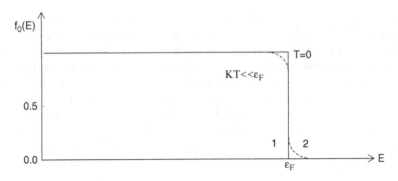

Fig. 2.1 Fermi distribution function. *1*: distribution function at absolute zero. *2*: distribution function at low temperature

movement of electrons will not be hindered and L is infinite. Electronic thermal conductivity could be infinitely large. In fact, the periodic lattice of metals is not complete. This is due to the thermal motion of atoms on the grid resulting from displacement from the equilibrium position, elastic distortion of the lattice caused by foreign atom, lattice dislocation, and fracture caused by grain boundaries. Therefore, electronic thermal conductivity is controlled by the scattering process and could not be infinitely large.

For lattice thermal conductivity, many factors, such as the nonresonance of the lattice wave, lattice defects, and the presence of impurity atoms, can impact the scattering mechanism. Therefore, the lattice thermal conducting process is a very complicated one. Because a heat carrier can only have a limited mean free path and the mean free path is in turn limited by various scattering mechanisms, for each heat carrier, its mean free path takes the following overlapping form:

$$\frac{1}{l} = \sum_{\alpha} \frac{1}{l_{\alpha}},$$

(2.5)

where α represents various scattering mechanisms.

In metals, the electron density is very high. At absolute zero, the energy of all electrons is below the Fermi energy level. As the temperature increases, the probability that an electron occupies an electronic state with the energy ε can be expressed as a Fermi distribution function curve (Fig. 2.1) and f_0 can be calculated using the following formula:

$$f_0 = \frac{1}{e^{\frac{\varepsilon - \varepsilon_F}{KT}} + 1},$$

(2.6)

where K is the Boltzmann constant.

It can be proved that at a normal temperature, or when $KT \ll \varepsilon_F$, the electron energy distribution in metals is similar to that at absolute zero. In other words, electrons are highly degenerate oriented. Therefore, the thermal conductivity characteristics of metals depend on the electron characteristics within an energy range of $\varepsilon_F \pm 0(KT)$. The main factors that affect the states of electrons are point defect electron scattering and electron–electron scattering.

It is obvious that thermal vibration and electron scattering are proportional to $\langle \varepsilon^2 \rangle$. Taking advantage of this point, the magnitude of *point defect electron scattering* can be estimated by the magnitude of thermal vibration at room temperature. In most solids, at room temperature $\langle \varepsilon^2 \rangle$ is approximately 0.01. The value of $\langle \varepsilon^2 \rangle$ is determined by atomic volume and elastic constants, but the difference is not large. Electrical resistivity and thermal resistivity at room temperature can be found to be

$$\rho_0 \approx (\rho_i)_{RT} c, W_0 T \approx (W_i)_{RT} c T_{RT}, \qquad (2.7)$$

where $(\rho_i)_{RT}$ and $(W_i)_{RT}$ are the intrinsic room temperature resistivity and thermal resistivity, respectively; c is the concentration of point defects (expressed as a percentage) and is about 300K.

Electron–electron scattering plays a fairly important role in determining the resistivity and thermal resistivity of metals with very high density such as transition metals. In this case, the electrical resistivity and the thermal resistivity have the following form:

$$\rho_{ee} = BT^2, W_{ee} = DT. \qquad (2.8)$$

In general, the electron–electron scattering effect at the low temperature limit (not a very low temperature) is important because for very pure samples, ρ_0 and w_0 do not play a major role. This is due to the relationship

$$\rho_i \propto T^5, W_i \propto T^2. \qquad (2.9)$$

In comparison, ρ_{ee} and w_{ee} could play more important roles than ρ_0 and w_0. Therefore, for the transition metals, in many cases ρ_{ee} and w_{ee} or the impact of the electron–electron scattering effect must be considered.

2.1.1.2 Mechanism of Metal Lattice Thermal Conductivity

The thermal motion of solid atoms in a crystal lattice contributes to thermal conductivity. In an insulator, the lattice thermal conductivity is almost the only mechanism and only at very high temperatures will thermal photons be of any significance. In metals, the main thermal conducting carriers are free electrons; in alloys, they are lattice thermal waves; and in semimetals and semiconductors, the magnitude of thermal conducting is often the same as that of electrons. Because the lattice vibration energy is quantized, the energy of a quantized lattice is called a quantum phonon with energy of ω, and ω is the angular frequency of the lattice wave.

We can see from the expression of thermal conductivity that lattice thermal conductivity is related to heat capacity, the average speed of lattice waves, and the mean free path of phonons.

The law of metal lattice thermal conductivity differs from that of insulators. In addition, electron–phonon scattering constitutes a difference in metal lattice thermal conductivity and insulator lattice thermal conductivity. For a complete crystal structure, a lattice wave has a frequency range, that is, from a low to a high frequency ω_m. ω_m is typically 10^{13} Hz. At low frequency, lattice waves can be regarded as elastic waves; at high frequency, due to the dispersion effect, their wavelength is similar to atomic spacing.

The interacting force between structural defects and atoms does not completely obey the law of elasticity (or Hooke's law). In fact, crystal lattice waves exchange energy with each other continuously, so each lattice wave has a finite mean free path l. Also, it is generally a function of frequency. Therefore, the lattice thermal conductivity can be expressed as

$$\lambda_g = \frac{1}{3}\int_0^{\omega_m} c(\omega)v_g l(\omega)d\omega, \tag{2.10}$$

where ω_g is the lattice wave group speed and $c(\omega)$ the specific heat. Because $c(\omega)$ is proportional to T^3 at very low temperatures and is independent of T at high temperatures, the lattice thermal conductivity λ_g depends mainly on the phonon mean free path. In what follows, we will discuss the law of metal lattice thermal conductivity from points of phonon–phonon scattering: from phonon–phonon scattering, electron–phonon scattering, and defect-phonon scattering.

• Phonon scattering on a phonon

At high temperatures, the under thermal strain, adjacent atoms are almost independent of one another and scattering is proportional to $<\varepsilon^2>$. That is, it is proportional to temperature. Therefore, the intrinsic mean free path is similar to the electron mean free path, namely:

$$l_i \propto \frac{1}{T}. \tag{2.11}$$

Since a lattice wave has a continuous spectrum range, every vibrating atomic position can be considered as a separate scattering source and its scattering effect is weak on a long wavelength lattice wave. And l_ω increases sharply with a decrease in ω. At low frequencies, l_i is proportional to $1/\omega^4$ and $c(\omega)$ is proportional to ω^2. As a result, at low temperatures, the lattice thermal conductivity (λ_g) integral diverges. Peierls made a rough estimate of the intrinsic thermal resistivity:

$$W_U = U\left(\frac{h}{K}\right)^3 \cdot \frac{\gamma^3}{Ma}\frac{T}{\Theta_D^3}, \tag{2.12}$$

where M is the atomic mass and a^3 the atomic volume; the coefficient U depends on the crystal structure with a typical value of 1/3; γ is a nonharmonic-vibration-related parameter, and Θ_D is the Debye temperature.

From the preceding formula it can be seen that the intrinsic thermal resistance is mainly determined by the Debye temperature. In general, the higher the Debye temperature, the higher the phonon thermal conductivity. According to Peierls's theory, the phonon process is divided into a forward process and a reverse process. At high temperatures, most of the phonon collisions are entail the reverse process. As in the previous analysis, the thermal resistance rate is proportional to the temperature, and the thermal conductivity is inversely proportional to the temperature.

At low temperatures, phonons reaching the reverse process change according to $e^{-\theta/bT}$. Therefore, thermal resistivity changes exponentially, that is, WU is proportional to $e^{-\theta/bT}$, where b is a constant related to the crystal structure. The variation agrees well with the experimental data.

- Electron–phonon scattering

In metals, electron–phonon scattering usually plays a major role at low temperatures. The effect of electron–phonon scattering is similar to the effect of lattice–electron scattering. These interactions limit the phonon mean free path. It is found that

$$\frac{1}{l_{pe}(\omega)} \propto \omega, \qquad (2.13)$$

where ω is the angular frequency of the lattice wave.

When λ_g is calculated according to Eq. 2.10, the effect of electron–phonon scattering must be taken into account for calculation of the mean free path. This scattering mechanism causes λ_g of metals to be smaller than λ_g of insulators with elastic properties. For the latter case, λ_g is proportional to T^2. As in the previous analysis, the high-temperature resistance of a nonharmonic vibration is dominant; that is, λ_g is proportional to $1/T_g$. Therefore, there is a maximum value of λ_g as the temperature changes from low to high. Usually, the maximum λ_g value of metals is smaller than that of insulators.

Equation 2.13 is based on the assumption that an electron has a long mean free path l', and it is longer than the wavelength of a lattice wave. That is, $l' > \lambda$. In this situation, phonons can interact with a single electron. In alloys, l' is limited. In the typical case, l' is about $100a$ (a^3 is the volume of atoms). If the concentration of impurity elements is 1%, then the wavelength of the lattice wave is

$$\lambda \approx \frac{1}{3} a \left(\frac{\Theta_D}{T} \right).$$

This condition of $l' > \lambda$ can still be satisfied. If the content of alloying elements increases, then there will be a situation where $l' << \lambda$. In this situation, phonons will no longer interact with single electrons; instead, the phonons behave as a whole electron gas. The scattering effect becomes

$$\frac{1}{l_{pe}} \propto \omega^2, \lambda_g \propto T. \tag{2.14}$$

In another case, even if the alloy composition is not high, at very low temperatures the lattice wave wavelength increases rapidly, and the effect of electron–phonon scattering is still obvious.

- Phonon scattering

There is a general rule about the effect of defect–phonon scattering. The main contribution to thermal resistance at low temperatures comes from a large area defects; the main contribution to thermal resistance at middle temperatures comes from point defects.

At very low temperatures, dislocation is usually the most important phonon scattering factor, the thermal resistance can be expressed as

$$\frac{1}{\lambda_g} = W_g = W_{ge} + W_{gd} \propto T^2, \tag{2.15}$$

where W_{ge} is the lattice thermal resistivity caused by electrons and W_{gd} is the lattice thermal resistance caused by dislocation.

When the density of dislocation reaches $10^{10}/cm^2$, then the value of W_{ge}, and W_{gd} are similar. Therefore, the lattice thermal conductivity of alloys is closely related to cold processing conditions, while the effect of cold processing upon electrical resistivity is not large. At low temperatures, the relationship between the phonon mean free path and frequency reflects the relationship between the lattice thermal conductivity and temperature. If there is the relationship

$$l(\omega) \propto \omega^{-n} \propto T^{-n} x^{-n},$$

then there is the relationship

$$\lambda_g \propto T^{3-n}, \tag{2.16}$$

where n denotes the form of the relevant defects; for point defects, $n=4$; for line defects, $n=3$; for sheet defects, $n=2$; for dislocation, $n=1$.

When the temperature is high and the concentration of point defects is high, strictly speaking, when $n=4$, Eq. 1.10 diverges. Therefore, the thermal resistance caused by point defects is often estimated using another method. That is, at low temperature, it is associated with electron scattering; at high temperatures, it is associated with nonharmonic vibrations. As the temperature is gradually increased from low to high, the Eq. 2.15 relationship will deviate more and more. Point defects will lower the maximum value of λ_g and broaden the temperature range of the maximum λ_g. At high temperatures, due to the presence of point defects, λ_g changes more slowly with T^{-1}. When the temperature and the concentration of point defects are high, the relationship between λ_g and the temperature will be

$$\lambda_g \propto [c(1-c)]^{-1/2} T^{-1/2}, \tag{2.17}$$

where c is the concentration of point defects.

2.1.1.3   Other Thermal Conducting Mechanisms

In addition to the two major heat transfer mechanisms—electronic and phonon thermal conducting mechanisms—there are a photon thermal conducting mechanism and a magnetic thermal conducting mechanism. If the thermal conducting medium is transparent at a particular wavelength range, then the thermal radiation can get through the media, hence the emergence of photon thermal conduction.

A magnon is a quantized unit of spin wave energy, defined in analogy with the phonon as a quantized lattice vibration. Interaction between magnons and spin waves develops into a new thermal conducting mechanism; it also becomes a thermal resistance mechanism in electronic and phonon thermal conduction. Various magnetic effects occur in rare earth materials. They have become a very important class of magnetic material. The magnetic effect of ferromagnetic and nearly ferromagnetic transition metals is also very important. The thermal conductivity of these materials is under continued research. No complete theory about their thermal conductivity has been developed yet.

The thermal conductivity of superconductors has unique characteristics. Below the superconducting transition temperature, some of the conduction electrons separated from the normal state are condensed into a state of zero entropy. They become so-called superfluid electrons and conduct current without resistance. The superfluid electrons do not conduct heat or scatter phonons. Under normal magnetic field constraints, thermal conductivity measurements can be performed under a normal state and a superconducting state.

2.1.2   Factors Affecting Thermal Conductivity

The thermal conductivity of materials is closely related to their material structure, density, temperature, and pressure. In general, to determine the precise thermal conductivity of a material, it must be measured by experimental apparatus. The factors that affect the thermal conductivity of materials still represent a very important topic. On the one hand, thermal conductivity can be used to identify the reliability of measured data; on the other hand, it can be used to make predictions about the properties of thermal conductivity of some materials. We will discuss four factors that can influence metal thermal conductivity.

1. Effect of atomic structure on thermal conductivity of metals

   Because free electrons play important roles in metal thermal conductivity, it is natural to link the thermal conductivity of metals with their atomic structure and with the periodic table of elements. If a suitable thermal conductivity unit is chosen, some regularity can be found. It has been found that, both in long periods and short periods, metals with one valence electron, such as the alkali metals, copper, silver, and gold, have the best conductivity. In the same period, thermal conductivity decreases from metals with one valence electron to metals with two valence electrons, such as from sodium to magnesium, or from copper to zinc. Some elements have only partially metallic characteristics, such as silicon, and their conductivity is low. In addition, some elements have a complex atomic structure, such as zirconium and titanium, and their conductivity is low.

2. Effect of chemical composition and structure on thermal conductivity of metals

   • The effect on thermal conductivity of adding a small amount of impurities

     The addition of impurities will cause the residual resistivity of the sample to increase, lowering the thermal conductivity. Studying the effect of impurities on the thermal conductivity of iron, it can be seen that the addition of foreign atoms will lower the thermal conductivity initially; as the amount of impurities increases, the proportion of thermal conductivity reduction decreases. In addition, if the atomic structure of the impurities is similar to that of the parent phase, the thermal conductivity will decrease less; an example is the case of adding cobalt and nickel. If the impurities' atomic structure and the structure of the parent phase are very different, then the thermal conductivity will decrease more; an example is the case of adding manganese and aluminum. The addition of the metalloid element silicon changes iron's thermal conductivity the most. When a variety of impurity atoms is added, the contributions to thermal resistance will be complex.

     There is another point to be made: the higher the thermal conductivity of the solvent (mother metal) element, the greater the impact of the impurity on the thermal conductivity of the parent metal.

   • Impact of orderness

     Slow cooling in an alloy or annealing at very low temperature may generate ordered alloys. In a typical body-centered cubic structure of an Fe–Co alloy, the iron atom percentage is 50%. Iron atoms are located in the center of a cube of cobalt atoms, whereas cobalt atoms are located in the center of a cube of iron atoms, forming an ordered alloy. This causes the mean free path of the conduction electrons to increase and their thermal conductivity is much greater than that of the disordered state.

3. Impact of porosity on thermal conductivity

   There are two types of metal materials that can be impacted by porosity: metal powder material (the metal is surrounded by air or gas) and pressed and sintered porous metal materials (where voids exist inside the solid metal). Their thermal conductivity can be estimated by Maxwell's equations. For powder material that is not sintered, if the powder is in a dispersed phase, then the thermal conductivity of the air is negligible relative to that of the powder and its conductivity can be obtained by

$$\lambda_p = \frac{\lambda_A (1 + 2\phi)}{1 - \phi},$$ (2.18)

where λ_p is the thermal conductivity of the powder, λ_A is the thermal conductivity of the air, and ϕ is the volume percentage of the metal powder. $\phi_A = (1 - \phi)$ is the volume percentage of the air. Then, the preceding equation can be written as

$$\lambda_p = \lambda_A \left(\frac{3}{\phi_A} - 2 \right).$$ (2.19)

This equation does not contain the thermal conductivity of a metal powder; the thermal conductivity of powder materials is directly proportional to the thermal conductivity of the surrounding air.

4. Effect of temperature on thermal conductivity
   Whether the thermal conductivity of a metal increases, decreases, or remains the same depends, with increasing temperature, on the temperature effect on the mean free path. In general, for pure metals, as the temperature increases, the mean free path decreases. The thermal conductivity change due to a temperature-induced mean free path reduction is more important than the direct effect of temperature on the thermal conductivity. Therefore, the thermal conductivity of pure metals generally decreases with increasing temperature. For alloys, because of the presence of foreign atoms, the effect of a change in the mean free path due to temperature is relatively small, and temperature plays a major role in the thermal conductivity of alloys; the thermal conductivity of alloys increases as the temperature increases.

2.1.3 Methods of Measuring Thermal Conductivity

Presently, there are two methods to determine thermal conductivity: steady-state and dynamic methods. The steady-state method is a classical method for determining the thermal conductivity of thermal insulation materials. It is still widely used and is based on the principle that the heat transfer rate is equal to the cooling rate at steady state. According to the Fourier one-dimensional steady-state heat conduction model, thermal conductivity is calculated by measuring the heat flux through a sample, the temperature difference at both ends of the sample, and the thickness of the sample.

Fig. 2.2 Flash method equipment for measuring thermal conductivity

The principle is simple and clear. This method has high accuracy, but the measurement time is long and it has strict requirements with respect to the environmental conditions. The steady-state method is divided into a heat flow meter method and a guarded hot plate method. The steady-state method is suitable for measurements of thermal conductivity at moderate temperatures. The guarded hot plate method has the highest accuracy, but measurements take a long time and the measurement process is complicated and expensive. The heat flow meter method is fast and accurate, but the measurement range is narrow. A dynamic method has been developed in recent decades. It is suitable for high-thermal-conductivity materials or in high-temperature conditions. The dynamic method requires a fixed-power heat source and records of sample temperature variations with respect to time. Thermal conductivity, thermal diffusivity, and heat capacity can be found from the sample temperature changes with respect time. The dynamic method is divided into a hot-wire method and a laser flash method. The measurement speed is high, but the accuracy is not (5%). The dynamic method is suitable for high-thermal-conductivity materials or in high-temperature conditions. A laser flash is generally used to measure the thermal conductivity of thermal management materials.

The laser flash method for thermal diffusivity measurement is based on the flash method first proposed by W.J. Parker in the United States in 1960. At that time, the sample was heated with a Xenon flash, and Xenon lights were later replaced by lasers. Seventy-five percent of the thermal diffusivity data is obtained from this measurement. The measuring principle is based on the assumptions that the circular sheet sample insulated from its surroundings is heated by a uniformly distributed instantaneous laser pulse, the temperature on the backside of the samples will rise, and the heat transfer is one-dimensional. Thermal diffusivity is calculated according to the theoretical model of thermal diffusivity. A set of measuring equipment is shown below (Fig. 2.2).

2.2 Coefficient of Thermal Expansion Requirement of Thermal Management Materials

In general, material will expand when the temperature increases and shrink when it cools. At the same temperature, different materials undergo different amounts of material expansion and contraction. Therefore, when two different materials are welded together, material expansion and contraction are restrained, resulting in thermal stress at the interface between the two materials. After a period of time and after repeated temperature cycling, thermal stress will cause material to bend, deform, or even crack. In the field of thermal management materials, die and package materials include Si, GaAs, InP, and Al_2O_3 or BeO and other semiconductor materials and ceramics. They are brittle materials. When they are welded together with a heat sink material, the difference between the CTEs is too large. After repeated temperature cycling, the chip or ceramic will warp and the welding joint could fail. Serious and complete failure or the cracking of ceramic materials and microchips could happen, affecting the life span and reliability of the components. In general, thermal management materials are required to have a CTE of less than $7 \times 10^{-6} K^{-1}$ and to have an optimal match of the CTE between thermal management materials and die materials to reduce interfacial thermal stress. The CTE match requirements of thermal management materials places limits on many good-thermal-conductivity materials with wide applications. For example, copper has good thermal conductivity, but unfortunately its CTE is up to $16.5 \times 10^{-6} K^{-1}$. The large CTE difference in dies and ceramics will cause the welding to deform and crack. Therefore, to meet the demands of high-power components, many researchers have developed functional composite thermal management materials, such as the second generation of the W–Cu and SiC–Al composites and the third generation of thermal management materials Diomand/Cu and C_f/Cu. These advanced materials have a low CTE and high thermal conductivity and their CTEs match those of the die materials. They represent the research and development direction for the future. Table 2.2 lists the CTEs of the second- and third-generation thermal management materials.

2.2.1 Definition of Coefficient of Thermal Expansion

The CTE is divided into the linear CTE and volume CTE. The linear CTE is defined as the average coefficient of linear expansion: when the temperature changes from t_1 to t_2, the length of the correspondingly the length changes from L_1 to L_2,

$$\bar{\alpha} = \frac{L_2 - L_1}{L_1(t_2 - t_1)} = \frac{\Delta L}{L_1 \Delta t}.$$

Table 2.2 Coefficients of thermal expansion of thermal management materials

Material	CTE (10^{-6}K^{-1}) 20°C	Material	CTE (10^{-6}K^{-1})
Si	4.1	W	4.6
Ge	5.5	Mo	5.4
SiC	4.0	Invar	0.4
GaAs	7.5	Kovar	4.2
InP	28	Al	22
96%Al$_2$O$_3$	6.3	Cu	17
AlN	4.2	Au96.85%-Si	12.3
BeO	6.4	Au80%-Sn	15.9
Epoxy	5.4	Au88%-Ge	13.4
		W85-Cu	6.9
		Mo85-Cu	6.7
		CMC1:1:1	8.9
		CPC1:4:1	7.8
		SiC70%-Al	7.0
		Si75%-Al	6.5
		C$_f$/Cu	6.5
		Diamond/Cu	

As Δt approaches zero, the preceding limit (in the case of constant pressure P) is defined as the differential linear expansion coefficient:

$$\alpha_t = \frac{1}{L}\left(\frac{\partial L}{\partial t}\right)_P.$$

Corresponding to this, when the temperature changes from t_1 to t_2, the volume of material changes from V_1 to V_2, and the average coefficient of volume expansion becomes

$$\bar{\beta} = \frac{V_2 - V_1}{V_1(t_2 - t_1)} = \frac{\Delta V}{V_1 \Delta t}.$$

Similarly, when Δt approaches zero, the preceding limit (in the case of constant pressure P) is defined as the differential volume expansion coefficient:

$$\beta_t = \frac{1}{V}\left(\frac{\partial V}{\partial t}\right)_P.$$

The differential coefficient of expansion is the expansion coefficient at a given point and at a certain temperature. For engineering applications, the average expansion coefficients are used most often.

2.2.2 Methods for Measuring the Coefficient of Thermal Expansion

Traditional methods for measuring the CTE of solid materials include push-rod dilatometer, optical lever dilatometer, the direct observation method, optical interferometry, X-ray method, the capacitance method, and others. Many of these methods were developed under certain conditions for specific research purposes; they cannot be used indiscriminately. Methods for measuring the CTE of composite materials can be divided into relative and absolute groups. The former group includes a volume method and a differential method; the latter group includes a telescope direct observation method and an interferometry method. The most recently developed laser method may be used more for measurements of composite materials.

The volume method is suitable for testing the composite volume expansion coefficient. The working principle is as follows. After the specimen is put into the liquid bath, the liquid height change is recorded along with the temperature change. The volume change and the volume CTE of the sample can be found after the effect of the liquid volume change itself due to the temperature change is eliminated.

The differential method is used based on the fact that a sample elongates as the temperature increases. The sample is connected to a micrometer and a gauge shows the total elongation amount of the specimen and the push rod. When the push-rod elongation amount is deducted, the specimen's amount of elongation is known. Thus, the linear expansion coefficient of the composite is obtained. This method is simple and reliable, and samples can be tested in a wide temperature range.

A high-sensitivity microtelescope direct observation method is used to observe the dimensional variation of test pieces due to temperature variation, resulting in a linear expansion coefficient. Its unique feature is that the upper temperature is high, currently up to 3,600°C. The working principle of an interferometer is introduced here. A change in specimen size will change the light optical path of one beam, so that the images rendered on a screen and the interference fringes will change. When enough fringe data are collected following the thermal expansion of the sample, the expansion length can be extrapolated using principles of physical optics. Thus, the linear CTE can be calculated. The greatest feature of this method is that even if the sample is thin, accurate results can be obtained; but the operation procedure is rather complicated and requires skilled specialists for testing. One piece of equipment for measuring the CTE is shown in Fig. 2.3.

2.3 Hermeticity Requirement of Thermal Management Materials

To ensure a chip has a good working environment, thermal management materials with good hermeticity are often used for hermetic packages. In a hermetic package, the required hermeticity is less than $1 \sim 5 \times 10^{-9}$ Pa.m^3/s. Hermeticity reflects the

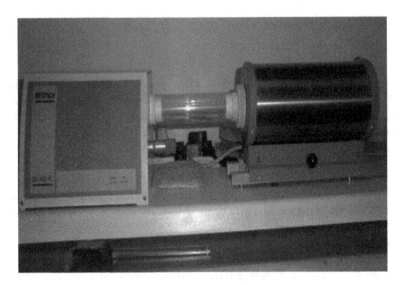

Fig. 2.3 DL402PC thermal conductivity measurement equipment

density of the material inside. The airtightness performance of many dense metals such as Kovar is good, and no extra testing is required; but for composite thermal management materials such as the refractory metal series W (Mo)–Cu and the aluminum matrix composite series SiC (Si)/Al, which are prepared by powder metallurgy, achieving good air tightness is one of the key technical process challenges. If the process is not controlled properly, problems related to tightness quality could occur, causing the entire batch of products to be scrapped. Inspection of airtightness during the production process is required.

In general, the hermeticity of thermal management materials is tested at the gross leak level and fine leak level. There are three types of gross leak test, namely, the bubble test, the weight gain test, and the dye penetrant test. In the bubble test, the device is immersed in a bath of indicator fluid at a temperature above the boiling point of the tracer fluid. The device must first be pressurized with the tracer fluid at a given temperature and pressure, which are determined in a way similar to fine leakage tests. The weight gain test involves cleaning and weighing the package before pressurizing it in a fluorocarbon tracer. The tracer material should have low viscosity and a low vapor pressure. The device is then dried and reweighed. The dye penetrant test utilizes a dye as a tracer; hence it is mostly used on transparent packages. Again, the device is pressurized in the tracer. The next stage, after washing, is to visually inspect (using a UV light) the interior to find dye inside of the cavity.

For a helium leak test, the package is placed in a pressurization tank, which is pressurized with helium at a given temperature for a given amount of time. The package is then removed and placed into a mass spectrometer so that the amount of helium coming out of the case can be observed. The test parameters are based on the internal volume of the package.

2.4 Other Performance Requirements of Thermal Management Materials

Electronic packaging materials must have high mechanical strength and good processing properties in order to be processed into a variety of complex-shape parts. In the aerospace field and in some portable electronic devices, electronic packaging materials must have lower density in order to minimize the weight of the devices. In addition, low cost is required for mass production.

2.5 Appearance Quality Requirements of Thermal Management Materials

2.5.1 Surface Layer of Nickel and Gold Plating Performance Requirements

To improve the reliability and stability of electronic components, component surfaces need to be gold plated. Before they are gold plated, the surfaces need to be nickel plated. For certain devices, nickel plating can facilitate silver and copper soldering. Therefore, thermal management materials should have very high-quality nickel and gold plating. For a nickel plating layer, a thickness of between 100 and 300 micron. is required. When they are heated at a high temperature of 850°C for 5 min, there should be no visible air bubbles under a 10× microscope; for a gold plating layer, a thickness of between 50 and 100 micron. is required. When the materials are heated at a high temperature of 350°C, there should be no visible air bubbles under a 10× microscope.

2.5.2 Quality Requirements of Surface Appearance

1. Surface roughness must be less than 0.8 μm, the lateral roughness less than 3.2 μm, the surface parallelness less than 0.02 mm, and the surface flatness less than 0.02 to 0.05 mm.
2. The surface scratch depth must be less than 0.0125 mm, the pinhole depth less than 0.025 mm, and the diameter less than 0.0127 mm.
3. There must be no visible burr on the surface; the side burr depth must be less than 0.02 mm.
4. There must be no oil spots, rust spots, or cracks on surfaces.

References

1. Chinese Society for Metals, Nonferrous Metals Society of China (1987) The physical properties of metallic materials handbook. Metallurgical Industry Press, Beijing, pp 299–320
2. Kun H (1988) Solid state physics. Higher Education Press, Hebei
3. Xide X, Junxin F (1961) Solid state physics (Volume). Shanghai Science and Technology Press, Shanghai
4. Zhang Ying et al (1997) Mater Rev 11(3): 52
5. German RM et al (1994) Powder Metallurgy Processing of Thermal Management Materials for Microelectronics Applcations. Int J of Powder Metall 30(2): 205
6. Bin Y, Renjie W, Zhang set (1994) Metal matrix composites for electronic packaging research and development. Mater Rev (3):64–66
7. Zweben C (1992) Metal-matrix composites for electronic packaging. JOM 44(7):P15–P24
8. Zweben C (1998) Advances in composite materials for thermal management in electronic packaging. JOM 50(6):47–51
9. Huang Qiang Gu, Ming-yuan JY (2000) Electronic packaging materials research. Mater Rev 14(9):28–32
10. The king of voice (2000) Multi-chip module (MCM) packaging technology. Microelectron 2(4):40–45
11. Liang SQ (2000) Epoxy resin in the packaging material of profiles. Thermosetting resin 15(1):47–51
12. Fan Y, Zhao Y-M (2001) Epoxy resins for electronic packaging research and development. Electron Process Technol 22(6):238–241
13. CF Legzdins etal (1997) MMCX - An expert system for metal matrix composite selection and design. Can Metall Quart 36(3): 177–178
14. Liang G (2003) Actively develop domestic microelectronics packaging industry. China Electron Bus 13(6):86–88
15. Strand SD (2005) Future technology in the global market. Power Systems World, 23–27 Oct 2005
16. Markoff J (2004) Intel's big shift after hitting technical wall, *New York Times*, 17 May 2004
17. Zweben C, Schmidt KA (1989) Advanced composite packaging materials. In: Electronic materials handbook. ASM International, Materials Park
18. Lasance CJM (2003) Problems with thermal interface material measurements: suggestions for improvement. Electron Cooling 9(4):22–29
19. Fleming TF, Levan CD, Riley WC (1995) Applications for ultra-high thermal conductivity fibers. In: Proceedings of the 1995 international electronic packaging conference, International Electronic Packaging Society, *San Diego*, pp 493–503
20. Norley J (2004) Natural graphite based materials for electronics cooling. In: Proceedings of the IMAPS advanced technology workshop on thermal management, *Long Beach*, 25–27 Oct 2004
21. Zweben C (2001) Electronic packaging: heat sink materials. Encycl Mater: Science Technol 3:2676–2683
22. Thaw J, Zemany J, Zweben C (1987) Metal matrix composites for microwave packaging components. Electronic Packaging and Production, pp 27–29
23. Zweben C (2006) Thermal materials solve power electronics challenges. Power Electron Technol 3(2):40–47

Chapter 3
Overview of Traditional Thermal Management Materials

Abstract In this chapter, we will present an overview of traditional thermal management materials. First, we will review the properties of Al_2O_3 dielectric materials and their applications in thick-film circuit substrate, thin-film circuit substrate, and multilayer substrate. Next, we will present the properties of Al_2O_3, BeO, AlN, SiC, and mullite and their manufacturing methods. In the following section, we will introduce traditional polymer-based thermal management materials, such as epoxy, silicone rubber, and polyimide. Finally, we will introduce pure metal or alloy traditional thermal management materials such as Cu, Al, W, Mo, and Kovar and their manufacturing methods.

Thermal management materials can be classified into several groups generally in accordance with the package structure, form, and material composition. On the basis of package structure, thermal management materials include mainly substrates, wiring, interlayer dielectric materials, and sealing materials. The substrates are generally divided into rigid and flexible boards. A flexible circuit board is light and thin. It may be suitable for portable electronics and wireless communications markets. Substrate metallization can wire chips on the substrates, and the wiring must have a lower resistivity and good solderability. The interlayer dielectrics are divided into organic (polymers) and inorganic (such as SiO_2, Si_3N_4, and glass) subgroups. They protect, insulate, and isolate the electronics to prevent signal distortion and other effects. Epoxy resin sealing materials account for 90% of the total current electronic sealing materials. Based on form, packaging can be divided into hermetic packaging and solid packaging. A hermetic package is one whose die cavity is filled with a certain atmosphere of space and isolated from the outside world; a solid package is one whose die surroundings are connected with the packaging structures. Based on material composition, materials can be divided into ceramic-based, plastic-based, and metal-based thermal management materials.

With the rapid development of electronic packaging technology, high-power electroniccomponents, and minimization have come more stringent requirements for packaging materials and more demand for the development of new thermal

G. Jiang et al., *Advanced Thermal Management Materials*,
DOI 10.1007/978-1-4614-1963-1_3, © Springer Science+Business Media New York 2013

management materials. In general, the development of thermal management materials shows the following characteristics:

- *Low density.* Because of the prevalence of portable electronic products, people are paying more and more attention to the issue of weight reduction; weight reduction for space applications is also important. Therefore, low-density materials such as aluminum, copper, and polymer materials will be widely used.
- *High thermal conductivity (TC), low expansion.* Because multichip packaging technology in the next few years may occupy an important position in the packaging field, and its density is increasing, the requirement for thermal conductivity of packaging materials has increased. Traditional electronic packaging materials cannot meet the requirement of high-density packaging. Therefore, many new electronic packaging materials with high thermal conductivity and low thermal expansion have been developed in recent years.
- *Integration.* As the number of ultrathin and ultrasmall electronic devices continues to increase, electronic packaging materials are also bound to become thinner and smaller. To reduce production costs and improve the reliability of packaging, it is expected that a combination of multicomponent, multifunctional, modular packages will be the future direction of development. New processing technology will have to be adopted to follow this trend.

Before we get into advanced thermal management materials, we will introduce the traditional thermal management materials.

3.1 Ceramic Matrix Conventional Thermal Management Materials

Ceramic-based thermal management electronic materials are commonly used, relative to plastic-based and metal-based thermal management materials. Their advantages are as follows:

1. Low dielectric constant, high-frequency performance;
2. Good insulation and high reliability;
3. High strength, good thermal stability;
4. Low thermal expansion coefficient, high thermal conductivity;
5. Good hermiticity, good chemical stability;
6. Good antimoisturization property, not prone to microcracking.

Microelectronic technology requires that the device package have a high density and fast heat dissipation and be lightweight, thin, fast, and inexpensive. Ceramic packages can largely meet these requirements, but at a higher cost. They are suited for advanced microelectronic device packages such as aerospace and military engineering applications, which require high-reliability, high-frequency, high-temperature, hermitic packages. They are also widely used in mobile communications, household appliances, automobiles, and other applications. Ceramic-based thermal management materials are produced mainly by a casting process. Casting originated in

1961. It is used in hybrid integrated circuits (HICs) and multichip module (MCM) ceramic packages. The United States, Japan, and other countries have developed many multilayer ceramic substrates. Now high-tech ceramics are widely used. The most commonly used ceramic thermal management materials include Al_2O_3, AlN, BeO, SiC, and mullite.

3.1.1 Al_2O_3 Thermal Management Materials

Al_2O_3 ceramics has a good overall performance and is currently the most mature thermal management material. Al_2O_3 raw materials are abundant. They are inexpensive, strong, and hard and show a good thermal shock tolerance, good insulation, chemical stability, and good adhesion to metal. Al_2O_3 ceramic substrates accounts for 90% of the total ceramic substrates. It is an indispensable material in the electronics industry. Depending on the concentration of Al_2O_3 content, it is divided into 99 ceramic, 95 ceramic, 90 ceramic, 85 ceramic, etc. For packaging applications, 99 porcelain is the most widely used material. It has 99.5% Al_2O_3 content. The overall performance of ceramics can be improved by increasing Al_2O_3 content, but the required sintering temperature increases with a corresponding increase in manufacturing costs. The sintering temperature and dielectric constant of Al_2O_3 mixed with Ag, AgPd, and other metal conductors or low-melting-point glass can be reduced. However, the CTE (7.2×10^{-6} K^{-1}) and dielectric constant (9.7) of Al_2O_3 ceramic substrates are higher than single-crystal S and its thermal conductivity (17 W/mK) is low, limiting its application in high-frequency, high power, ultra-large-scale integrated circuits (ICs).

Al_2O_3 thermal management materials are mainly used as a thick-film circuit substrate, thin-film circuit substrate, and multilayer substrate.

3.1.1.1 Thick-film Substrate

The requirements of a thick-film substrate are that it must have a flat and smooth surface finish for screen printing, withstand sudden increases and sudden decreases in temperature for thick-film sintering and dip soldering processes, possess good mechanical strength to withstand stress, and show good thermal performance. In addition, it should have good electrical insulation and high-frequency characteristics and good dimensional accuracy, and be inexpensive. Thick-film circuit substrates are found in approximately 90–97% of Al_2O_3 ceramics.

3.1.1.2 Thin-film Circuit Substrate

In addition to the requirements for thick-film substrates, the performance of a thin-film circuit substrate is also very sensitive to surface roughness, and surface finish requirements are higher. Surface roughness should be below 0.025 μm or better.

Most Al_2O_3 thick-film ceramic substrates have a surface roughness of 0.13–0.20 μm. Under normal circumstances, Al_2O_3 with smaller grains and higher purity has a better surface finish.

3.1.1.3   Multilayer Substrate Integrated Circuits

Currently, Al_2O_3 ceramic multilayer substrates are mostly used to improve the integration of 3 day ICs. Multilayer substrate technology takes the following three forms:

1. Thick-film printing multilayer
   On a sintered Al_2O_3 ceramic substrate, a multilayer is formed by laminating Au, Cu, and other conductive pastes with a low-dielectric paste. The multilayers are sintered in 850–900°C. This substrate is inexpensive, easy to replace, and easily forms the desired impedance. But to prevent the occurrence of a fault, up to 5–6 layers are used and the line width cannot be too narrow.
2. Preform multilayer
   Through-holes are formed on presintered Al_2O_3 preforms. Circuits made of W, Mo, and other high-melting-point metal conductive pastes are printed. After printing, presintered preforms are laminated together. The laminated multilayers are sintered in a reducing atmosphere.
3. Preform printing multilayer
   On the performs, W or Mo conductive paste and Al_2O_3 dielectric paste are printed alternatively and then sintered together. The performance of a multilayer substrate made in this way falls between the two methods above. Its main features are that the design can be changed easily for a variety of products and the production cycle is short.

To improve the thermal conductivity of Al_2O_3 ceramics, Larson and others use diamond-coated Al_2O_3 ceramic substrates to improve the thermal performance. They deposited a layer of composite thin film containing a diamond phase and SiO_2 using plasma CVD technology on top of the Al_2O_3 ceramic substrate; the reaction gases contained C, H, Si, and O elements in the gas mixture. The results showed that the thermal conductivity of the diamond-coated Al_2O3 ceramic substrate was significantly improved; the temperature gradient under the same conditions as the general Al_2O_3 ceramic substrates was reduced by approximately 50%.

3.1.2   BeO Thermal Management Materials

BeO ceramics have a piezoelectric property, photochemical property, high strength, low dielectric constant, low dielectric loss, high adaptability in packaging technology, and other characteristics. A BeO ceramic substrate has a high thermal conductivity,

Table 3.1 BeO ceramic material properties with different BeO purities

BeO content/%				
Property	95	98	99	99.5
Density/(g/cm^3)	2.8	2.85	2.9	2.9
Bending strength/MPa	186	186	186	186
TC/(W/mK)	201	205	243	255
Bulk resistivity 20°C/(Ω.m)	>10^5	>10^5	>10^{15}	>10^{15}
30°C/(Ω.m)	10^{11}	10^{15}	10^{15}	10^{15}
Permittivity 1 MHz below	6.5	6.5	6.8	7.1
Dielectric losses 1 MHz below	7×10^{-4}	1×10^{-4}	5×10^{-4}	2×10^{-4}

close to that of metals. Its thermal performance is significantly better than that of Al_2O_3 ceramics. It is also a good insulating material and, therefore, a widely used material for high-power electronic components. BeO ceramics with a purity greater than 99% and with a 99% relative density have a thermal conductivity at room temperature of up to 350 W/mK; but the thermal conductivity of BeO ceramics containing other ingredients with different structures would drop dramatically. Therefore, the purity requirement for highly thermal conductive BeO ceramics is very demanding, generally more than 99%. In addition, the thermal conductivity of BeO ceramics drops significantly with increasing temperature. At 1,000°C, thermal conductivity decreases to 10% of the thermal conductivity at room temperature.

By BeO content, BeO ceramics are divided into 95, 98, and 99% BeO ceramic. Its main properties are shown in Table 3.1. BeO ceramic thermal conductivity is nine times that of Al_2O_3 ceramics. Unfortunately, BeO ceramic dust is highly toxic and expensive. Protective measures must be taken in production; high processing temperatures are required. Therefore, the production cost of BeO substrates is high; in addition, the production process pollutes the environment, limiting its production and applications. At present, BeO substrates are mainly used in high-frequency, high-power electronic device cooling, aerospace equipment, and other applications. A BeO sheet with metal coating has been used in aircraft control systems and automotive ignition devices. Pacific Microelectronics makes low-cost BeO ceramic substrates with BeO coarse particles.

A BeO wurtzite structure is shown in Fig. 3.1. The crystal structure is a hexagonal close pack (hcp) array of Be^{2+} with O^{2-} in half the tetrahedral sites.

The dark gray balls represent beryllium atoms, and the light gray balls represent oxygen atoms

The BeO ceramic production process follows the usual porcelain process: preparation, molding, and sintering. Because the toxicity of BeO powder is strong, direct contact with BeO should be avoided during production; special attention must be paid to prevent BeO dust and air pollution. A BeO ceramic is a high-refractory oxide with a melting point of 2,570°C. It has strong covalent bonds. The sintering temperature of pure BeO ceramics is 1,900°C or more. To reduce the sintering temperature, Al_2O_3 and MgO, and other agents are often added. However, thermal conductivity can drop by 15% with an additional 1% SiO_2 in the BeO ceramic.

Fig. 3.1 Organizational
structure of BeO

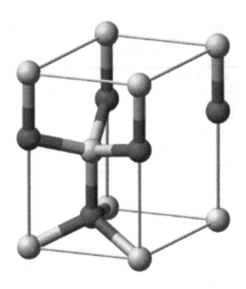

3.1.3   AlN Thermal Management Materials

AlN is a promising thermal management material with excellent electrical and
thermal performance. Compared with Al_2O_3, AlN has higher thermal conductivity
and its CTE matches well with that of Si. Its dielectric constant is lower, suitable for
high-power, multilead, large chips. AlN has high hardness and is suitable for harsh
environments; AlN thin substrates can be made to meet the needs of a variety of
packaging substrates, suitable for production of ultrafast, ultra-large-scale IC
substrates.

AlN has a diamond crystal structure similar to the wurtzite-type structure for
covalent nitrides with a theoretical density of 3.26 g/cm^3 and decomposition temperature of 2,790°C and a theoretical thermal conductivity of up to 319 W/mK. Due
to the strong covalent bonding in AlN, sintering is difficult, especially at atmospheric pressure sintering. To promote densification, sintering additives must be
used. The sintering additives include rare earth oxides and alkaline earth oxides
such as B_2O_3, Y_2O_3, La_2O_3, and CaO. In the sintering process, these additives react
with oxygen impurities in AlN powder, producing additive-Al-O compound oxides
and promoting densification of AlN.

Therefore, the research on AlN ceramics has focused on the reduction of sintering temperature and improvement of thermal conductivity. Liang and others add
CaF_2 and YF_3 sintering agents, at 1,750°C, N_2 atmosphere sintering, a thermal conductivity of 180 W/mK of AlN ceramics was obtained. Li and others obtained an
AlN ceramic with a relative density of 96.3% at 50 GPa, 1,300°C sintering for
50 min. Therefore, to obtain AlN ceramic with a high purity and high density, the

Table 3.2 Comparison of translucent AlN ceramic and Al_2O_3 and BeO ceramic performance

Properties	AlN	Al_2O_3	BeO
Purity/%	>99.5	>96	>99.5
Hardness HV/GPa	11.76	24.5	11.76
Density/(g.cm^{-3})	2.36	3.98	2.9
Bending strength/MPa	392	294	186
TC/(W/mK)	140	20	255
CTE (20–400°C)/10^{-6} K^{-1}	4.4	7.2	8.0
Bulk resistivity 20°C/(Ω.m)	10^{14}	10^{14}	>10^{15}
Permittivity 1 MHz and below	8.9	9.4	7.1
Dielectric losses 1 MHz and below	58×10^{-4}	4×10^{-4}	2×10^{-4}

oxygen content in the raw materials must be low, the grain size small, and the sintering temperature above 1,600°C with an N_2 presence. It is even better to have high temperature and high pressure at the same time. By adding sintering agents such as CaO, Y_2O_3, and others, not only does a liquid phase develop, but the densification and mechanical strength are improved, the oxygen content can be reduced, and the thermal conductivity is improved. The purity of AlN is increased with the presence of N_2. The preparation process of AlN ceramics is complicated and costly, and large-scale production and application have yet to be developed.

"SHAPAL" AlN is a translucent ceramic thermal management material made by a reduction nitriding method. It is made with high-purity, fine-grained AlN powder, and to the powder is added 1.0 wt.% of a CaO sintering agent. The preforms can be made by a cast-molding method or dry compression method; finally, sintering is performed in hot pressing or pressureless conditions. Fully dense translucent AlN ceramic thermal management materials have been obtained with a density of 99.5%. Their properties, along with the properties of Al_2O_3 and the BeO performance of thermal management materials, are shown in Table 3.2.

3.1.4   SiC Thermal Management Material

SiC compounds have strong covalent bonds and a diamond-type structure. There are 75 variants of such compounds. The main variants are α-SiC, 6H-SiC, 4H-SiC, 15R-SiC, and β-SiC, where H represents the hexagonal structure and R represent the rhombohedral structure. Numbers before H and R represent the number of layers in a repeated cycle along the c-axis. α-SiC is a hexagonal crystal with a high-temperature stable structure. β-SiC is a cubic crystal and has a low-temperature stable structure. Starting from 2,100°C, β-SiC undergoes transformations into α-SiC. SiC ceramics have no melting-point temperature. At 1 atmospheric pressure and a temperature of 2,830±40°C SiC decomposes into silicon and carbon.

SiC ceramics have a high thermal conductivity (at room temperature in the range of 100–490 W/mK) and a low CTE. Its CTE matches well that of Si. It has good

Table 3.3 Physical properties of SiC material

Properties	SiC ceramic		
	Dip silicon method	Pressure sintering	CVD
Density/(g.cm^{-3})	3.15	3.15	2.9–3.19
Bending strength	460	350–430	~500
Hardness HV	2,800	2,800	3,000–4,000
Modulus/GPa	410	410	≥480
TC/(W/mK)	80.9	91.96	83.73
CTE (20°C)/10^{-6} K^{-1}	4.4	4.8	4.8

electrical insulation properties and high mechanical strength. The SiC ceramic sintering process is difficult, and it is necessary to add a small amount of Be or Al oxide as a sintering agent to increase the density. Studies have shown that Be, B, and Al and its compounds are effective additives. Density up to 98% of pure SiC ceramics can be obtained, but the dielectric constant of SiC is high and the dielectric strength is low, limiting high-frequency applications used only in low-density packaging.

SiC ceramics can be prepared by reactive sintering and densification sintering with additives. Reactive sintering includes a high-temperature recrystallization method and SiC-C system green-body-forming silicide sintering methods. Another way to make SiC ceramics is to make SiC-C preforms first. Then they are heated to 1,650°C. The preforms are dipped into melted silicon. SiC prepared in such a way is very dense, but the residual silicon dramatically reduces the mechanical strength in a range of 1,200–1,400°C. The physical properties of SiC ceramics are shown in Table 3.3.

3.1.5   Mullite Thermal Management Materials

Mullite made of Al_2O_3 and SiO_2 is the only stable binary compound at normal pressure. Its chemical formula is $3Al_2O_3 \cdot 2SiO_2$. There is little natural mullite. It can be made by sintering or synthesized by an electrical melting method. Mullite's mechanical strength and thermal conductivity are lower than those of Al_2O_3. It has a low dielectric constant, low CTE, and high hardness, in addition to good thermal and chemical stability. Mullite is an alternative material to Al_2O_3 and has been extensively developed. The dielectric constant, CTE, substrate deformation, and stress can be reduced by adding MgO.

3.1.6   Multilayer Cofired Ceramics

Multilayer cofired ceramic substrates are monolithic substrates made by lamination, hot pressing, binder removal, and sintering processes. Due to multilayer, higher wiring density and shorter interconnect lengths, packing densities and signal transmission speeds have increased, meeting the electronic system miniaturization requirements for high reliability, high efficiency, and high power. Therefore, they are widely used in a wide range of applications. There are two kinds of multilayer cofiring process: a high-temperature cofiring and a low-temperature cofiring process. High-temperature cofiring is mainly suited for Al_2O_3, AlN, and other materials. Generally, the temperature is in the 1,650–1,850°C range, and W, Mo, Mn, and other refractory metal wire materials are used for conducting wires. The advantages of high-temperature cofiring are that it produces high mechanical strength, high thermal conductivity, high wiring density, and high chemical stability. For high-temperature cofired ceramic (HTCC) circuit interconnect substrate, W and Mo have a high resistivity, and the circuit resistance loss is high. As the frequency and speed of ultra-large-scale ICs increased and the density of the packaging increased due to the miniaturization of electronic devices, low-temperature cofired ceramics (LTCC) came into being. LTCC is divided into three categories: glass-ceramic systems, glass systems, and nonceramic glass systems. LTCC sintering temperatures are mainly between 800 and 900°C. The raw ceramic material is mixed with an organic binder, and the preform is prepared by a casting process. Through-holes are punched in the preforms, and the through-hole metallization and printed conductor pattern form electrical connections between the layers. After the lamination, hot pressing, and binder removal processes are finished, the multilayer is sintered at a low temperature below 980°C. The main difference between LTCC and HTCC substrates is that the ceramic powder ingredients and metallization materials are different. The LTCC sintering process is easy to control. It has high-dimensional accuracy and its production costs are low. It can be cofired with Cu, Ag, Ag2Pd, and Au. It has a low dielectric constant (6.5–7.2, 1 MHz), low dielectric loss, and low CTE [$(5.3–5.6) \times 10^{-6}\,K^{-1}$], good bending strength (250 MPa), and high-frequency performance. LTCC is an ideal material for high-frequency applications. LTCC is currently widely used in the military, aerospace, automotive, electronics, medical, and other fields. LTCC will be the leading electronic ceramic packaging technology in the future.

3.2 Traditional Plastic-based Thermal Management Materials

For plastic-based thermal management materials, the cost is low and the manufacturing process is simple. Among electronic thermal management materials, it is the most widely used and fastest-growing material. It is one of the most important classes of materials to achieve miniaturization of electronic products; it is lightweight and low-cost thermal management material. Plastic-based thermal management materials are not dense enough; the ion content is high, and the temperature reliability is not good enough. As the properties of the raw materials and the formulations are improved, these issues will be gradually resolved. At present, the problem that needs to be addressed is the fact that the CTE of plastic-based thermal management materials does not match that of silicon wafers. The ideal plastic-based thermal management material should have the following properties:

1. High material purity, very few ionic impurities;
2. Good adhesion of the lead frame with the device;
3. Low water absorption and low water vapor transmission rate;
4. Low internal stress and shrinkage;
5. Low CTE, high thermal conductivity;
6. Fast formation and curing, good mold release;
7. Low viscosity, fast filling, flashless;
8. Flame retardant.

Most plastic-based thermal management materials are thermosetting plastics, including epoxy, phenols, polyester, and silicone (silicone plastic). The commonly used plastics are epoxy, silicone rubber, polyimide, and others.

Epoxy molding compounds (EMCs) are composed of phenolic epoxy resin, phenol resin, filling (SiO_2), mold release agents, curing agents, dyes, etc. EMCs have excellent adhesion and electrical insulation, high strength, good heat tolerance and chemical resistance, low water absorption, and good molding characteristics. EMC-based plastic packaging accounts for more than 90% of packaging in the industry. According to a report, by mixing the negative thermal expansion material ZrW_2O_8 powder by a certain percentage with the E51 epoxy resin by ultrasonic treatment, ZrW_2O_8 powder can be uniformly dispersed in the epoxy matrix. As the fraction of ZrW_2O_8 mass is increased, the CTE is reduced and the glass transition temperature and tensile and bending strength are increased. Rimdusit developed new polymeric systems based on the ternary mixture of benzoxazine, epoxy, and phenolic novolac resins. A glass transition temperature as high as 170°C and considerable thermal stability at 5% weight loss up to 370°C can be obtained from these systems. Phenolic novolac resin acts mainly as an initiator for these ternary systems, while low melt viscosity, flexibility, and improved crosslink density of the materials are attributed to the epoxy fraction. Polybenzoxazine imparts thermal curability, mechanical properties, and low water uptake to ternary systems. The materials exhibit promising characteristics suitable for application as underfilling encapsulation and other highly filled systems.

Silicone rubber has good heat aging resistance, UV aging resistance, and insulating properties. It is mainly used in semiconductor and LED packaging. According to one report, a colorless and transparent silicone thermal management material was obtained by mixing silicone, organic silicone oil, and a reaction catalyst. It can be used for high-power white LED packaging. A light transmission rate of 98% and white LED luminous flux of up to 42–65 lm were obtained. When epoxy resin is used as lens material, anti-aging properties are obviously inadequate. It is incompatible with the heat management material interface and dramatically reduces the life of a LED. Silicone rubber is a material that shows good interface compatibility and anti-aging properties. High refractive index silicone rubber materials have become a priority in development and production.

Polyimide can withstand temperatures of up to 350–450°C. It has good insulation, good dielectric properties, and good resistance to organic solvents and moisture. It has been widely used in the semiconductor and microelectronics industry. A polyimide layer is mainly used as a passivation layer, stress buffer, protective coating, interlayer dielectric material, or liquid crystal orientation film. It is especially suitable for flexible circuit substrates.

3.3   Pure Metal or Alloy Traditional Thermal Management Materials

Traditionally, both pure metals and alloys, such as Cu, Al, W, Mo, and Kovar, have been the most widely used heat sink materials. The thermal properties of some pure metals and alloys are shown in Table 3.4.

3.3.1   Pure Copper and Aluminum Metal Thermal Management Materials

Copper and aluminum are inexpensive and have good thermal conductivity. Copper has a thermal conductivity of 400 W/mK at room temperature, whereas aluminum has a thermal conductivity of 240 W/mK at room temperature, second only to silver. From the perspective of heat transfer, both copper and aluminum are ideal as thermal management materials and can be used in applications requiring high thermal conductivity and high conductance. Initially, copper and aluminum were used in power devices. Currently, aluminum is the main heat spreading substrate material in LEDs. However, copper and aluminum both have high CTEs—16.5×10^{-6} K^{-1} in the case of copper and 23×10^{-6} K^{-1} for aluminum, which are much higher than those of matching ceramics and Si chips. When they are bonded to a rigid ceramic substrate, a large thermal stress is produced, affecting the life span and reliability of the electronic components. To reduce the stress on the ceramic substrate, designers replace one single substrate with several smaller substrates with separate wiring. In addition, the mechanical properties of copper and aluminum are poor. High-temperature-induced softening in the packaging and sealing results in permanent deformation or even cracking. Therefore, to improve the annealing temperature of copper, a small amount of Al_2O_3, zirconium, and silicon oxides are added, enhancing the high-temperature mechanical strength of copper by a dispersion strengthening effect. As a result, the annealing temperature of oxygen-free high conductivity is increased from 320 to 420°C, the reduction in thermal conductivity and electrical conductivity is not significant. Dispersion-enhanced, oxygen-free copper made by SCM Metal Products'

Table 3.4 Thermal properties of pure metals and alloys

Material	Density(/cm³)	TC(W/mK)	CTE(10^{-6} K^{-1})	Modulus (GPa)
Cu	8.96	400	16.5	110
Al	2.7	230	23	70
W	19.3	170	4.6	400
Mo	10.2	140	5.4	340
Kovar	8.2	17	4.2	138
#10 steel	7.8	49.8	12.6	207

Glidcop contains 0.3% Al_2O_3. Its thermal conductivity reaches 365 W/mK. This material is widely used in metal packaging.

Copper thermal management materials are generally oxygen-free or dispersion enhanced oxygen-free. Their purity can reach 99.97% or more. Impurities such as Fe, Ni, etc. have a strong negative impact on the thermal conductivity of copper, but the effect of Zr, Cr, and other elements is not significant. Therefore, such impurities are used to strengthen the oxygen-free copper, and high-strength, dispersion-enhanced, oxygen-free copper is obtained. In addition, among impurities, the most harmful element is oxygen. If a high level of oxygen is present in copper, hydrogen embrittlement phenomena will occur, resulting in cracking or deformation of the copper. Therefore, the oxygen content in oxygen-free copper must be less than 20 ppm, or even less than 5 ppm.

3.3.2 Refractory Metals Tungsten and Molybdenum as Thermal Management Materials

Tungsten and molybdenum are VIB group elements in the periodic table. Their important physical properties and mechanical properties at room temperature are shown in Table 3.5.

As can be seen from Table 3.5, tungsten and molybdenum have high melting points, high elastic moduli, high-temperature strength, and high density. Tungsten has the highest density in metals, followed by molybdenum. The CTEs of tungsten and molybdenum are relatively small. They match well those of Kovar, ceramics, and Si chips. In addition, their thermal conductivities are very high. They are used as the base of power electronic components and in many middle- and high-power density metal packages. Figure 3.2 shows components made by a molybdenum material used in power electronics. But the processing capability and solderability of tungsten and molybdenum are poor. They are more expensive and not suitable for large quantity use. In addition, tungsten and molybdenum have recrystallization brittleness and their BDTT temperatures are relatively high.

Tungsten is very stable in a dry environment at room temperature, but it will slowly oxidize in moist air. At 400°C, tungsten oxidize in a minor way producing blue oxide ($WO_{2.9}$). When the temperature exceeds 400°C, tungsten rapidly oxidizes to brown tungsten oxide (WO_2) or yellow tungsten oxide (WO_3). Below 400°C molybdenum oxidizes, generating a dense adhesive oxide film. But when the temperature exceeds 725°C, the surface will be catastrophically oxidized and the yellowish molybdenum oxide will sublimate. Tungsten and molybdenum are both high-melting-point metals and can be prepared by a powder metallurgy method and a melting method. But the electron beam melting method is expensive, and these two metals are generally prepared by a powder method. Specific processes are as follows (Figs. 3.3 and 3.4).

Table 3.5 Physical and mechanical properties of tungsten and molybdenum at room temperature

Property	Mo	W
Crystal structure	b.c.c	b.c.c
Density(g/cm^3)	10.2	19.3
Melting point/°C	2,620±20	3,380±20
Boiling point/°C	4,804	5,930
CTE/10^{-6} K^{-1}	5.3	4.6
CTEt (20°C)/(W/mK)	140	170
Recrystalline point/°C(1 h annealing)	800	1,150
Modulus/GPa	320–360	390–410
Poisson ratio	0.3	0.3
Vickers hardness		
Recrystalline (minimum)	150	350
Process state (maximum)	500	650
BDTT/°C	−20	+400
Electronic work function/eV	4.55	4.2

Fig. 3.2 Molybdenum materials used for power semiconductor devices

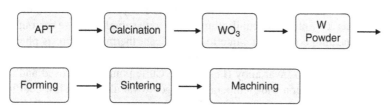

Fig. 3.3 Preparation diagram of tungsten

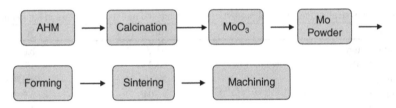

Fig. 3.4 Preparation diagram of molybdenum

Fig. 3.5 Kovar metal package

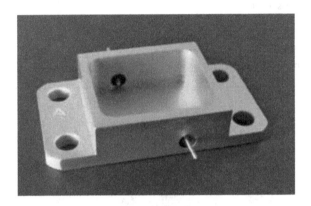

3.3.3 Kovar Alloy

Kovar, a low expansion alloy, has the chemical formula Fe-29Ni-17Co. It is composed of 29% nickel and 17% cobalt. Its CTE is close to those of Si, GaAs, Al_2O_3, BeO, and AlN. It has good weldability and workability. In a range of 20~450°C, it can be sealed with hard borosilicate glass and 95% Al_2O_3 ceramic. It is widely used in low-power-density metal packages, as shown in Fig. 3.5, a Kovar metal shell. However, due to its low thermal conductivity, high resistivity, and high density, its applications are greatly limited.

Kovar's physical properties are shown in Table 3.6.

Kovar alloy must have a single-phase γ solid solution microstructure and no trace of martensitic when cooled to −70°C to prevent devices from bursting upon volume expansion due to martensitic transformation expansion. At room temperature Kovar has α + γ phases at equilibrium, close to the single γ-phase region; but at the usual annealing cooling rate, due to a significant thermal hysteresis phenomenon, a single γ-phase structure can be obtained.

By adding cobalt to Invar alloy (Fe–Ni), the Curie point is improved and a low CTE is maintained over a wide temperature range. Kovar alloy has a low coefficient of thermal expansion due to the "Invar" antiexpansion phenomenon. Below the Curie point, volume expansion and contraction result from a magnetostrictive effect.

Table 3.6 Physical
properties of Kovar alloy

Property	Kovar
Hardness HB	140–160
Density/(g.cm^{-3})	8.3
Yield strength/MPa	343
Tensile strength/MPa	500–618
Elongation δ/%	35–37
TC/(W/mK)	18
CTE (20–300°C)/10^{-6} K^{-1}	4.7–5.5
(20–400°C)/10^{-6} K^{-1}	4.6–5.2
(20–500°C)/10^{-6} K^{-1}	5.9–6.4
Bulk resistivity 20°C/(μΩ.cm)	49
Magnetoconductivity	800

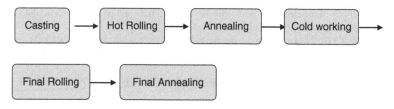

Fig. 3.6 Kovar preparation flow chart

But when the temperature exceeds the Curie point, the "Invar effect" disappears, and the CTE of Kovar increases significantly.

A Kovar alloy preparation flow chart is shown in Fig. 3.6.

3.3.4   #10 Steel

Thermal steel has a conductivity of 49.8 W/mK and a CTE of 12.6×10^{-6} K^{-1}. Its CTE is three times that of Kovar alloy. Its CTE does not match that of Kovar. The #10 steel cannot be sealed by a matching glass seal; it must be sealed by a compression seal. The seal is achieved by the cooling contraction of the steel that is greater than the glass shrinkage because the compressive strength of glass is much greater than the tensile strength of glass.

The physical properties of #10 steel are shown in Table 3.7.

Table 3.7 Physical
properties of #10 steel

Property	#10 Steel
C/%	0.07–0.14
Density/(g.cm^{-3})	7.8
Yield strength/MPa	205
Tensile strength/MPa	335
Elongation δ/%	31
TC/(W/mK)	48.6
CTE (20–300°C)/10^{-6} K^{-1}	12.6

References

1. Harper CA (2005) Electronic materials and processes handbook, 3rd edn. McGraw-Hill, New York
2. Zhou L (2006) Microelectronic device packaging materials and packaging technology package. Chemical Industry Press, Beijing
3. Carl ZM et al (1992) Matrix composites for electronic packaging. JOM 44(7):15–23
4. Carl Z (1998) Advances in composite materials for thermal management in electronic packaging. JOM 50(6):47–51
5. Junjun C, Yuepeng F, Mingbo T (2008) The latest progress in microelectronic packaging materials. Semicond Technol 33(3):185–189
6. Tian M (2003) Electronic packaging engineering. Tsinghua University Press, Beijing
7. Hodson TL (1995) AlN steps up, takes the heat and delivers. Elect Packag Prod 35(7):26–30
8. Liang Q, Zhou HP, Fu RL et al (2003) Thermal conductivity of AlN ceramics sintered with CaF2 and YF3. Ceram Int 29:893–897
9. Larson SE, Slaby J (2004) Comparison of various substrate technologies under steady state and transient conditions. Integr Electron Syst Sector 2:648–653
10. Vaed K, Florkey J, Akbar SA et al (2004) An additive micromolding approach for the development of micromachined ceramic substrates for RF applications. J Microelectron Mech Syst 30(13):514–521
11. Li XL, Ma HA, Zuo GH et al (2007) Low temperature sintering of high density aluminium nitride ceramics without additives at high pressure. Scipta Mater 56(12):1015–1018
12. Zaosheng Z, Zhengya L, Zhiwu C (2008) Ceramic substrates for electronic packaging materials. Mater Guide 22(11):16–20
13. Gorbunova MA, Shein IR (2007) Electronic and magnetic properties of beryllium oxide with 3d impurities from first principles calculations. Phys B Condens Matter 1:47–52
14. Zhou Q (2009) LTCC technology development and trend. Chinese Electron J l(4): 25–26
15. Guifang X, Xiaonong C, Wei X et al (2008) Epoxy/ZrW2O8 preparation and properties of packaging materials. J Jiangsu Univ 25(3):223–226
16. Wenying H (2008) Performance improvement of electronic packaging materials. Appl Science Technol 22(16):15–20

Chapter 4
Development of Advanced Thermal Management Materials

Abstract In this chapter, we will present the development of advanced thermal management materials. First, we will introduce one popular classification of thermal management materials. Next, we will present thermal management materials with an Al–Cu matrix and particle-enhanced materials such as W, Mo, SiC, AlN, BeO, Si, and others with a low coefficient of expansion. The materials covered include W-Cu, Mo-Cu, AlSiC, Cu/SiC, Cu/Si, and negative thermal expansion materials. In the following section, we will introduce fiber-reinforced thermal management materials such as boron fibers, carbon fibers, Al_2O_3 fibers, and SiC fibers. The development of a Cu–C_f composite and aluminum graphite materials is presented. Finally, copper/molybdenum/copper (CMC), copper/molybdenum–copper/copper (CPC), and Cu/Invar/Cu (CIC) materials are introduced.

4.1 Introduction

Dr. Carl Zweben, Advanced Thermal Materials Consultant from Devon, PA, one of the most well-known experts in the field, summarized the latest development in heat sink materials in *Power Electronics Technology*. He classified the thermal management materials into three generations. First-generation thermal management materials include copper, aluminum, glass fiber-reinforced polymer, copper/tungsten (Cu/W), copper/molybdenum (Cu/Mo), copper-Invar-copper (Cu/I/Cu), and copper-molybdenum-copper (Cu/Mo/Cu) and (Cu/Cu-Mo/Cu). Their properties are listed in Table 4.1. Tables 4.2 and 4.3 present the properties of several dozen selected second-generation and third-generation high-performance materials, respectively. Table 4.3 includes diamond made by chemical vapor deposition (CVD) for reference. For anisotropic materials, inplane isotropic and through-thickness thermal conductivity (k) values are presented. The absolute and specific thermal conductivities of the advanced materials in Table 4.2, and especially in Table 4.3, are significantly higher than those of the traditional materials in Table 4.1. One of the second-generation

G. Jiang et al., *Advanced Thermal Management Materials*,
DOI 10.1007/978-1-4614-1963-1_4, © Springer Science+Business Media New York 2013

Table 4.1 Properties of first-generation heat sink materials (From [1])

Reinforcement	Matrix	Thermal Conductivity (W/mK)	CTE (ppm/K)	Density (g/cm^3)	Specific thermal conductivity (W/mK)
–	Aluminum	218	23	2.7	81
–	Copper	400	17	8.9	45
–	Invar	11	1.3	8.1	1.4
–	Kovar	17	5.9	8.3	2
–	Cu/Invar/Cu	164	8.4	8.4	20
–	Cu/Mo/Cu	182	6	9.9	18
–	Cu/Mo–Cu/Cu	245–280	6–10	9.4	26–30
–	Titanium	7.2	9.5	4.4	1.6
Copper	Tungsten	157–190	5.7–8.3	15–17	9–13
Copper	Molybdenum	184–197	7.0–7.1	9.9–10.0	18–20
–	Sn63/Pb37	50	25	8.4	6
–	Epoxy	1.7	54	1.2	1.4
E-glass fibers	Epoxy	0.16–0.26	11–20	2.1	0.1

Table 4.2 Properties of second-generation heat sink materials (From [1])

Reinforcement	Matrix	In-plane thermal conductivity (W/mK)	Through-thickness thermal conductivity (W/mK)	In-plane CTE (ppm/K)	Density (g/cm³)	Specific in-plane thermal conductivity (W/mK)
Natural graphite	Epoxy	370	6.5	-2.4	1.94	190
Continuous carbon fibers	Polymer	330	10	-1	1.8	183
Discontinuous carbon fibers	Copper	300	200	6.5–9.5	6.8	44
SiC particles	Copper	320	320	7–10.9	6.6	48
Continuous carbon fibers	SiC	370	38	2.5	2.2	170
Carbon foam	Copper	350	350	7.4	5.7	61

Table 4.3 Properties of third-generation heat sink materials (From [1])

Reinforcement	Matrix	In-plane thermal conductivity (W/mK)	Through-thickness thermal conductivity (W/mK)	In-plane CTE (ppm/K)	Density (g/cm³)	Specific in-plane thermal conductivity (W/mK)
–	CVD Diamond	1,100–1,800	1,100–1,800	1–2	3.52	310–510
–	HOPG	1,300–1,700	10–25	–1	2.3	565–740
–	Natural graphite	150–500	6–10	–	–	–
Continuous carbon fibers	Copper	400–420	200	0.5–16	5.3–8.2	49–79
Continuous carbon fibers	Carbon	400	40	–1	1.9	210
Graphite flake	Aluminum	400–600	80–110	4.5–5.0	2.3	174–260
Diamond particles	Aluminum	550–600	550–600	7.0–7.5	3.1	177–194
Diamond and SiC particles	Aluminum	575	575	5.5	–	–
Diamond particles	Copper	600–1,200	600–1,200	5.8	5.9	102–203
Diamond particles	Cobalt	>600	>600	3	4.12	>145
Diamond particles	Silver	400 to >600	400 to >600	5.8	5.8	69 to >103
Diamond particles	Magnesium	550	550	8	–	–
Diamond particles	Silicon	525	525	4.5	–	–
Diamond particles	SiC	600	600	1.8	3.3	182

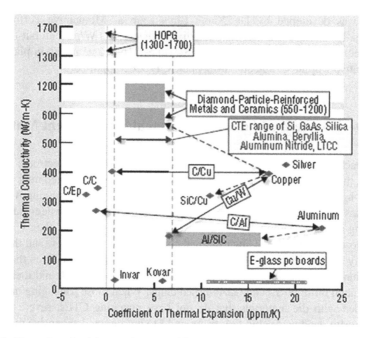

Fig. 4.1 Thermal conductivity as a function of CTE

thermal management materials, silicon carbide particle-reinforced aluminum, commonly called Al/SiC in the packaging industry, is a metal-matrix composite (MMC) that was first used in microelectronic and optoelectronic packaging by Dr. Zweben and his colleagues at GE in the early 1980s.

Figure 4.1, which plots thermal conductivity as a function of CTE, compares traditional and advanced thermal materials. Ideal materials have high thermal conductivities and CTEs that match those of semiconductors and ceramics like Si, GaAs, alumina, aluminum nitride, and low-temperature cofired ceramics (LTCCs). As the figure shows, by combining matrices of metals, ceramics, and carbon with thermally conductive reinforcements like special carbon fibers (abbreviated C), SiC particles, and diamond particles, it is possible to create new materials with high thermal conductivities and a wide range of CTEs.

Materials presented include monolithic metals, highly oriented pyrolytic graphite (HOPG), and a number of composites. The composites include carbon-fiber-reinforced carbon (C/C), carbon-fiber-reinforced epoxy (C/Ep), carbon-fiber-reinforced copper (C/Cu), silicon-carbide-particle-reinforced copper (SiC/Cu), and traditional Cu-W. HOPG, also called thermal pyrolytic graphite and annealed pyrolytic graphite by various manufacturers, and diamond-particle-reinforced metals and ceramics have the highest thermal conductivities. Most recently, thermal pyrolytic graphite, a unique form of pyrolytic graphite manufactured from the thermal decomposition of hydrocarbon gas in a high-temperature chemical vapor deposition reactor, was developed by Dr. Xiang Liu's team at Momentive Performance Materials. Its in-plane thermal conductivity reached 1,500 W/mK, an out-of-plane thermal conductivity of less than 20 W/mK, a density of 2.2 g/cm^3, an in-plane CTE of 0-negative 1 ppm/°C, and an out-of-plane CTE of 25 ppm/°C.

4.2 Development of Advanced Thermal Management Materials

As the packing density of integrated circuits and the power of electronic devices increase, the requirements for electronic packaging materials are becoming increasingly stringent. As we saw in Chap. 3, traditional electronic packaging materials such as Invar, Kovar, Al, Cu, W, Mo, etc. cannot meet the growing demands of the packaging industry. Although Invar and Kovar have low CTEs and their CTEs can match those of die materials, their thermal conductivities are too low and they cannot satisfy the requirements of high-density packaging technology. Even though Al and Cu have high thermal conductivities, their CTEs are too high and their CTEs cannot match those of Si- and GaAs-based chips. When those packaging materials are bonded with die materials that do not have matching CTEs, severe thermal stresses result, leading to chip failure. W and Mo are rare and expensive materials, and their processing performances are poor and production costs high. While their CTEs are low, they do not form good bonding with BeO or Al_2O_3 substrates due to a CTE mismatch. Therefore, the need to develop packaging materials with combined properties of low CTE, low density, high thermal conductivity, suitable strength, and low production costs is well justified. In general, it is difficult to satisfy the demanding packaging requirements with a single material; only a composite material can fulfill that role. The performances of composite materials can be adjusted, taking full advantage of the benefits of a single material.

The development of new composite materials for thermal management is focused on two types of composite materials: polymer-matrix composites and metal-matrix composites. The most obvious advantage of polymer-matrix composites is their low density. As was previously mentioned, epoxy resin is widely used mainly due to its good adhesive properties, corrosion resistance, and electrical performance. However, epoxy material is brittle and has low strength and poor thermal conductivity. Researchers have tried various methods to modify properties of epoxy resins. One way is to add a large amount of SiO_2 particles to the epoxy resin. If the appropriate processing technology is applied with coupling agents, the volume fraction of SiO_2 particles can greatly improve the heat resistance of epoxy resin, resulting in not only an increase in heat resistance, but also a reduction in the material's moisture absorption and low CTE. A low CTE is the primary consideration in selecting packaging materials. As the thermal conductivity of polymers is generally poor and the density of electronic packaging is increasing, it is critically important to improve the thermal conductivity of polymers, so that heat can be removed in a timely fashion. To significantly improve the thermal conductivity of epoxy materials, simply relying on the thermal conductivity of the epoxy resin itself is not enough; other materials with good thermal conductivity must be considered. The decision on which of these materials to introduce must be based not only on whether or not the material's thermal conductivity and thermal expansion match, but also on their impact on the environment and production costs. Boron nitride, aluminum nitride, and beryllium oxide all have high thermal conductivities, but beryllium oxide is highly toxic. Boron nitride can be added in limited amounts. The manufacturing costs of aluminum nitride are high, and its performance is greatly affected by the production process. Magnesium oxide and alumina have achieved large-scale industrial production. They are nontoxic, inexpensive to produce, and have a low CTE. Nanomaterials can be used as epoxy resin modifiers. There are many dangling bonds on the surface of nanomaterials, and the odds that unpaired atoms will form a physical or chemical bond are high. Therefore, the interfacial bonding between the particles and the matrix is enhanced. The reinforced, toughened material can bear greater loads. The overall performance of composite materials can be improved effectively by adding a small volume fraction of inorganic particles dispersed in epoxy resin, and the composite is made with very little volume fraction of inorganic particles. Therefore, if the nanomaterial is used effectively in the epoxy matrix, then the polymer-matrix composite could have high strength, high toughness, and high heat resistance.

MMCs have long been the subject of study, and there exists a complete theoretical description of this class of composite materials. MMCs have good overall properties in terms of thermal conductivity, electromagnetic shielding, and production cost. In addition, they are widely used. It is generally agreed that MMCs represent one of the most important directions for the future development of thermal management materials. MMCs have a high thermal conductivity, high electrical conductivity, and low CTE. This book focuses on MMCs. Depending on the matrix base, MMCs can be classified mainly into Cu-based and Al-based composites. Depending on the composite structure, they can be classified into particle-reinforced, fiber-reinforced, and "sandwich"-reinforced structures.

4.3 Particle-enhanced Thermal Management Materials

The particles used for the enhancement of thermal management materials are W, Mo, SiC, AlN, BeO, Si, and others with low CTEs. Matrices are mainly composed of Al and Cu. What distinguishes particle-enhanced thermal management materials from conventional particle-reinforced composites is that for conventional composites, the volume fraction of reinforcement is very small, whereas for particle-enhanced thermal management materials, the volume fraction of reinforcement is very large. To lower the CTE of composite materials to form good CTE matches with Si, GaAs, and other die materials, a large volume of low-CTE particles is incorporated because Al and Cu have large CTEs. For example, in a W–Cu composite, the tungsten content is more than 80%; in Al–SiC composites, the SiC content is in the

Fig. 4.2 Typical W-Cu heat sinks

range of 55–70 vol.%. Their CTE is in the range of 6.0–7.3×10^{-6} K^{-1}, their thermal conductivity is in the range of 170–220 W/mK. We will introduce the most commonly used MMCs in this chapter.

4.3.1   W-Cu Thermal Management Material

Tungsten has a very high melting point, high density, and low CTE. Copper is a good electrical and thermal conductor. The combination of both tungsten and copper (i.e., copper–tungsten composite) enjoys the low CTE of tungsten and high thermal conductivity of copper. At the same time, the CTE and electrical conductivity can be adjusted by varying the copper-to-tungsten ratio. Copper tungsten is widely used as electrodes and as heat sinks for microelectronics packaging. Figure 4.2 shows some typical W-Cu heat sinks.

Copper tungsten's CTE can be designed to match the CTE from ceramics, semiconductor chips, and other metals. For microelectronics packaging applications, W-Cu has some unique requirements for CTE, thermal conductivity, and hermeticity. Consequently, the manufacturing process has some unique challenges:

1. Meeting CTE, thermal conductivity, and hermeticity specifications;
2. Meeting all dimensional requirements;
3. Meeting surface finish requirements (like Ni or Au plating).

There are many grades of W-Cu materials currently being used as heat sinks. Their typical technical properties are shown in Table 4.4.

Table 4.4 Typical properties of heat sink grade W-Cu (Courtesy Torrey Hills Technologies)

Name	Property				
	Density (g/cm³)	CTE ($10^{-6}K^{-1}$)	TC (W/mK)	Young's modulus (GPa)	Hardness (HV10)
W90Cu	17.0	6.5	190–200	330	300
W88Cu	16.9	6.8	190–200	320	290
W85Cu	16.3	7.0	200–210	310	280
W80Cu	15.6	8.0	210–220	280	260

Table 4.5 Typical properties of heat sink grade Mo-Cu (Courtesy Torrey Hills Technologies)

Physical property	Mo50Cu50	Mo60Cu40	Mo70Cu30	Mo80Cu20	Mo85Cu15
Composition (wt%)	50% Mo Cu: balance	60% Mo Cu: balance	70% Mo Cu: balance	80% Mo Cu: balance	85% Mo Cu: balance
Density at 20°C (g/cm³)	9.5	9.6	9.7	9.9	10.0
CTE at 20°C (ppm/K)	9.9	9.5	7.5	7.2	6.8
Thermal conductivity (W/mK)	250	215	195	175	165
Specific heat at 100°C (J/kgK)	323	310	301	–	275
Specific electrical resistance at 20°C (μ$\square$m)	0.028	–	0.37	–	–
Young's modulus at 20°C (GPa)	172	–	225	21	248
Flexural strength (MPa)			–	1,103	1,138
Vicker's hardness (HV 10)	150	–	170	–	–

4.3.2   Mo-Cu Thermal Management Material

The Mo–Cu composite shows combinational properties such as having high electrical and thermal conductivities and low CTE, being nonmagnetic, having good high-temperature performance, and many other advantages. They have a bright future in the field of electronic packaging applications. Compared with traditional packaging materials, they have a high thermal conductivity and their CTEs are closely matched to those of die materials. Their CTEs can be adjusted, which is a strong advantage; compared with AlSiC composite materials, various methods of mechanical cutting processing can be performed. Compared with W-Cu materials, molybdenum in Mo-Cu has a relatively low density; the use of Mo-Cu can reduce the weight of packaging materials. Because of this unique advantage, Mo-Cu is a preferred choice in aerospace, instrumentation, and portable equipment industries.

There are many grades of Mo-Cu materials currently being used as heat sinks. Their typical technical properties are shown in Table 4.5.

Table 4.6 Density of W-Cu and Mo-Cu (Courtesy Torrey Hills Technologies)

Mo-Cu	Mo-15Cu	Mo-20Cu	Mo-30Cu	Mo-40Cu	Mo-50Cu
$\rho(g/cm^3)$	10.0	9.9	9.8	9.66	9.54
W-Cu	W-10Cu	W-15Cu	W-20Cu	W-30Cu	W-40Cu
$\rho(g/cm^3)$	17.3	16.42	15.66	14.31	13.17

4.3.3 Comparison Between Mo-Cu and W-Cu Packaging Materials

A Mo-Cu composite material was developed in the 1960s. It exhibits good electrical conductivity, good thermal conductivity, good corrosion resistance, good machinability, and adjustable CTE. It is widely used as a heat sink in the microelectronics packaging, vacuum electrical contactors, aerospace, and mining industries.

4.3.3.1 Density

Like tungsten, molybdenum is a refractory metal with a very high melting point. At the same time, it has a low CTE and high thermal conductivity, making it a good material for heat sinks. Among heat sink materials its CTE is one of the closest to that of silicon. In addition, molybdenum is widely used as a carrier (tabs) for die attachments. Since the density of Mo is much lower than that of tungsten, it is desired in many weight-sensitive applications like aerospace and portable electronics. In addition, the Mo-Cu composite is easier to machine.

Table 4.6 shows the density comparison between W-Cu and Mo-Cu.

As shown in Table 4.6, with the same copper content, Mo-Cu is about 50–75% lighter than W-Cu. Once again this property is desirable in weight-sensitive applications.

4.3.3.2 Coefficient of Thermal Expansion

The CTE changes with the temperature. Figure 4.3 shows a CTE comparison among 95% alumina, W-15Cu, and W-15Mo. In a range of 100–800°C, the CTE of Mo-15Cu matches that of 95% alumina very well (and is better than that of W-15Cu). This translates to reduced interface residual stress and improved component reliability.

4.3.3.3 Specific Heat

Specific heat (Cp) is another important property for heat sink materials. A high Cp is desirable since a heat sink material with a higher Cp can absorb more heat itself and reacts better with the peak and valley of the waste heat. Since Cp is not sensitive

Fig. 4.3 CTE comparison of W-15Cu and Mo-15Cu (Courtesy Torrey Hills Technologies)

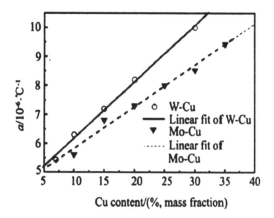

Fig. 4.4 Specific heat comparison between W-Cu and Mo-Cu (Courtesy Torrey Hills Technologies)

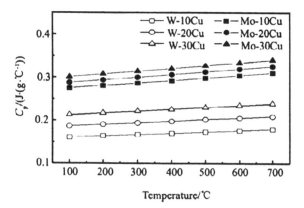

to microstructures, W-Cu's and Mo-Cu's Cp in a temperature range of 100–700°C was estimated using the theory of combination and is summarized in Fig. 4.4. When the heat sink mass and area are fixed, Mo-Cu's higher Cp gives it better heat dissipation.

4.3.3.4 Copper Wettability

Figure 4.5 shows the wetting angles θ of molten copper on tungsten and molybdenum plates. Overall, molten copper wets tungsten better than molybdenum. Like tungsten, molybdenum does not alloy with copper easily. To sinter Mo and Cu, some sintering aids like Ni, Co, Fe, and Pd can be used. Among these sintering aids, Ni works the best; it helps lower copper's wetting angles on molybdenum and promotes densification.

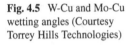

Fig. 4.5 W-Cu and Mo-Cu wetting angles (Courtesy Torrey Hills Technologies)

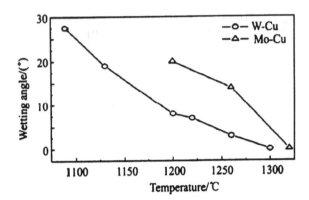

4.3.4   AlSiC and AlSi Thermal Management Materials

An MMC is a composite material with at least two constituent parts, one being a metal. The other material may be a different metal or another material such as a ceramic. MMCs are made by dispersing a reinforcing material into a metal matrix. For microelectronics packaging applications, we are interested in particle-reinforced MMC materials. They normally use SiC, AlN, BeO, Al_2O_3, and Si, low-CTE particles to disperse in a metal matrix like aluminum.

Aluminum and aluminum alloys like aluminum-silicon (eutectic, hypo, and hyper), aluminum-copper, aluminum-silicon-copper, aluminum-magnesium, aluminum-silicon-magnesium, aluminum-silicon, and magnesium-copper form the base materials for composites. They are available in the open market covered by IS, BS, and ASTM specifications. Properties of the composite can be made to suit the required specifications by the proper selection of the base alloy and the percentage of graphite to be added to sink materials.

For packaging heat sink materials, only AlSiC, Al-Si, and Al-graphite are used commercially and will be discussed in detail here.

4.3.4.1   AlSiC Thermal Management Materials

AlSiC is an MMC composed of silicon carbide particles and an aluminum alloy. As shown in the grayscale micrograph in Fig. 4.6, the silicon carbide particles (dark gray) are surrounded by a continuous aluminum matrix (off-white). Typically, AlSiC composites are isotropic and do not have an aluminum skin, which can often result in nonisotropic behavior, i.e., bimetallic deformation.

There are three main challenges to fabricating AlSiC materials.

1. Aluminum does not wet SiC
   Molten aluminum does not wet SiC. During the manufacturing process, it is very important to modify the surface of SiC to make sure the AlSiC interface forms a strong bond.

Fig. 4.6 SEM micrographs of AlSiC (Courtesy Thermal Transfer Composites, used with permission)

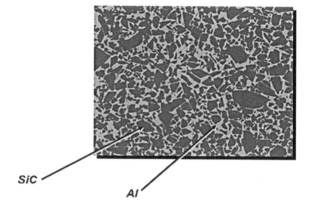

SiC

Al

2. AlSiC is hard to machine

   SiC is a very hard material and is often used as a grinding medium. Machining AlSiC to various heat sink shapes is very hard. EDM, waterjet, and diamond grinding are commonly used.

3. Plating

   Since SiC is not electrically conductive, it is not possible to use a conventional electrical plating method to deposit any Ni or Au plating. AlSiC material is typically electrolessly plated first. The adhesion layer between SiC and the metal is critical to prevent any separation (like delamination and blisters). The AlSiC composite will be covered in detail in a subsequent chapter.

4.3.4.2   Aluminum Silicon Thermal Management Materials

Hypereutectic aluminum silicon alloys containing from approximately 16 to 19% by weight of silicon possess good wear-resistant properties achieved by precipitated primary silicon crystals. The conventional aluminum silicon alloy usually contains a substantial amount of copper, generally in the range of 4.0–5.0%. Because of the high proportion of copper, the alloy has a relatively wide solidification temperature range in the neighborhood of about 136–167°C, which severely detracts from the castability of the alloy. The copper also reduces the corrosion resistance of the alloy in saltwater environments and thus prevents its use for marine engines.

Hypereutectic aluminum silicon alloys have been deemed significant for use by the casting industry as well as the automotive industry. This principally results from the potential that such alloys hold for providing good wear resistance along with the conventional advantages derived from aluminum castings. Furthermore, it is well recognized that the casting characteristics of hypereutectic alloys are very good.

A leader in developing AlSi materials is Sandvik Materials Technology (Fig. 4.7). Table 4.7 lists the properties of typical AlSi materials.

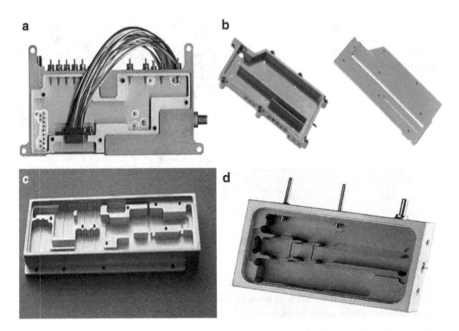

Fig. 4.7 Typical AlSi parts. (**a**) AlSi transmit-receive module. (**b**) Housing for radar circuitry. (**c**) Hybrid electronics packages. (**d**) Power amplifier housing (Photos courtesy Sandvik Materials Technology)

Table 4.7 Properties of typical AlSi materials

Composition	Al-27%Si	Al-42%Si	Al-50%Si	Al-60%Si	Al-70%Si
CTE at 25°C (ppm/K)	15.3	12.2	11.4	9.1	7.2
Thermal conductivity (W/mK)	177.4	160	149	129.4	120
Specific heat (J/gK)	0.85	0.82	0.79	0.78	0.78
Young's modulus (GPa)	91.8	101.9	121.4	123.5	129.2
Density (g/cm³)	2.6	2.55	2.51	2.46	2.42

4.3.5   Cu/SiC and Cu/Si Thermal Management Materials

Just as SiC and Si form composites with Al, so SiC, SiC, and Si form composites with Cu. Because the thermal conductivity of Cu is better than that of Al, in theory, the thermal conductivities of SiC/Cu and Si/Cu should be better than those of AlSiC and AlSi, with a slight increase in density. In the following sections, we will introduce SiC/Cu and Si/Cu.

Fig. 4.8 Effect of SiC
content on CTE and thermal
conductivity of Cu/SiC

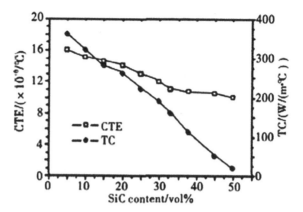

4.3.5.1 SiC/Cu Thermal Management Materials

The thermophysical properties of SiC/Cu thermal management materials are affected by the SiC fraction, particle size, shape, and other factors (Fig. 4.8). With the increase of SiC content, the thermal conductivity of Cu/SiC decreases significantly, resulting in a reduced CTE. According to a study by Zhu and others, in an SiC particle size range of 10–20 μm, the Cu/SiC CTE is not very different. However, if the SiC particle size exceeds 63 μm, then the CTE will be higher than that of the former two. The CTE increases significantly when the temperature increases. The effect of SiC particle size on Cu/SiC thermal conductivity is very weak. SiC in Cu/SiC packaging materials mainly takes on the form of SiC particles, fiber, and 3D interpenetrating network type. The thermal conductivity of particulate Cu/SiC falls in a range of 250–325 W/(m°C) and the thermal expansion coefficient in a range of $(8.0–12.5) \times 10^{-6}/°C$. Schubert produced particle-type Cu/SiC packaging materials with thermal conductivities [W/(m°C)] of 222 and 288 and with corresponding CTEs of 14.5, 10.6, and 11.2 $(\times 10^{-6}/°C)$. Yih et al. prepared fiber-type Cu/SiC packaging materials with an SiC fiber content of 30% (volume fraction), a thermal conductivity of 190 W/(m°C), and a CTE of $12 \times 10^{-6}/°C$. Xing studied a 3D2SiC netlinking system with diffused Cu phase and prepared a composite packaging material with a high fraction of Cu phase and a low fraction of SiC phase.

4.3.5.2 Cu/Si Thermal Management Materials

Cu/Si not only has the many advantages of Cu/SiC; it also has easy machinability, so the Cu/Si has a greater advantage. However, Cu/Si packaging materials progress slowly and few results are available mainly due to the appearance of Cu_2Si and the processing of Cu/Si packaging materials.

According to the rule of mixture (ROM), under ideal conditions, the thermal conductivity and CTE of Cu/Si material should vary with the Si content (Fig. 4.9).

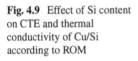

Fig. 4.9 Effect of Si content on CTE and thermal conductivity of Cu/Si according to ROM

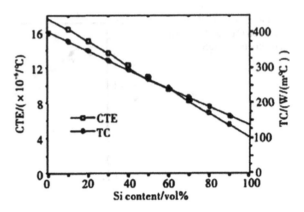

Table 4.8 Properties of Cu/Si composite materials

Si content (%)	50	60	70	80
CT (W/mK)	47	35	14	10
CTE ($\times 10^{-6}$ K^{-1})	15.2	14.0	10.1	8.5

When the CTE is less than $10 \times 10^{-6}/°C$, the thermal conductivity of Cu/Si is greater than 250 W/(m°C), and this correlation can be used to design the thermal conductivity and CTE of Cu/Si packaging materials. Lee and others prepared Cu/Si packaging materials using a hot-pressing method at 500°C, 400 Mpa, and a holding time of 10 min. Their properties are listed in Table 4.8.

The actual properties of Cu/Si composites are relatively poor, much lower than the Cu/Si theoretical values, mainly because of the interaction of copper with silicon, leading to the loss of high thermal conductivity of the copper phase. The diffusion rate of Cu in Si is fast; a solid solution of copper in silicon takes the form of Cu_2Si. At the same time, many Cu_2Si compounds have been developed. According to the Cu_2Si binary phase diagram, at 200°C, Cu_2Si can be generated. Copper silicon compounds include Cu_3Si, Cu_5Si, and so on. As in the case of Cu/SiC, a diffusion barrier layer can be employed to prevent a Cu/Si interface reaction.

4.3.6 Negative Thermal Expansion Materials, Thermal Management Materials

This type of thermal management material has a low CTE, which is negative. When it is incorporated into composite materials, a large volume fraction of high thermal conductivity material such as Cu or Al can be added; in this way the thermal performance of thermal management materials can be significantly improved. The CTEs of composite materials made with low volume fraction of negative thermal expansion materials could be used to match those of Si and GaAs, and their CT loss is low. Commonly used negative thermal expansion materials include ZrW_2O_8, $Zr_2(WO_4)(PO_4)_2$, ZrV_2O_7, other compounds, and shape-memory alloys like TiNi. At room temperature and atmospheric pressure, there are two ZrW_2O_8 cubic phases. When the temperature is below 423 K, it takes on the α-ZrW2O8 phase; when the temperature exceeds 423 K, it takes on the β-ZrW_2O_8 phase. Under high pressure, it takes on the γ-ZrW_2O_8 orthorhombic phase. The CTEs of three phases are all negative but vary in their values (-8.7, -4.9, -1.0) ($\times 10^{-6}/°C$) for α, β, and γ phases. For Cu/60% (volume fraction) α-ZrW_2O_8, it is estimated that the calculated CTE value is in the range of $(1.1–4.5) \times 10^{-6}/°C$. In one experiment, Cu/75% (volume fraction) ZrW_2O_8 composite material was prepared. The ZrW_2O_8 powder surfaces were pre-copper plated, and then the powder was prepared using a hot isostatic pressing method (100 MPa, 500°C/3 h). The measured CTE fell in the range of $(4–5) \times 10^{-6}/°C$. It was estimated that the CTE of TiNi alloy in the range of 25–100°C was $-21 \times 10^{-6}/°C$. The prepared Cu/TiNi composite had a CTE value of $4 \times 10^{-6}/°C$.

When the negative thermal expansion materials are processed, interfacial reactions and diffusions take place, resulting in a failure to obtain the desired negative thermal expansion material. Verdon et al. prepared Cu/33% (volume fraction) ZrW_2O_8 composite material preforms using a Cu/ZrW_2O_8 powder mixture under pressure at 300 MPa and then hot-pressed at 103 MPa, 600°C, for 3 h. However, ZrW_2O_8 was decomposed into WO_3 and other compounds. In addition, at room temperature there is a big difference in CTE values between ZrW_2O_8 and Cu. Therefore, when the composite is cooled to room temperature from the sintering temperature, very large residual stress is produced due to the CTE mismatch. As a result, the production cost of Cu/ZrW_2O_8 negative thermal expansion materials is high.

4.4   Fiber-reinforced Thermal Management Materials

This type of thermal management material still uses highthermal-conductivity Cu and Al as its base; low-thermal-expansion materials such as boron fibers, Al_2O_3 fibers, SiC fibers, and carbon fiber as reinforcements. The most widely used are the carbon fibers, and it is now possible to obtain high-thermal-conductivity electronic packaging materials. Composite materials with small CTEs and high thermal conductivity have been prepared using C/Al, C/u, SiC_f/Al. The thermal conductivity of some C/Cu composites reaches 400 W/mK. However, the bond strength between carbon fibers and a matrix is weak; C/Al composites undergo serious intergranular corrosion and interfacial reactions. In addition, carbon fibers have large anisotropy. To avoid anisotropy of composite materials, it is necessary to make the carbon fibers into mesh, spiral, and skew mesh forms. Thus, the manufacturing process is rather difficult and costly and makes it difficult to realize large-scale mass production. Such materials are used only in military, aerospace, and other important applications.

4.4.1   Introduction to Reinforced Fibers

4.4.1.1   Boron Fiber

Boron fiber was first used for high-performance reinforced fiber composite materials. It exhibits high elastic modulus, good wettability between fiber and metal base, and little reactivity between fiber and metal base. In addition, the fiber diameter is large. Generally, boron on heated tungsten is deposited by CVD in a hydrogen atmosphere. The boron fiber composite material breaks easily in the longitudinal direction because of its larger diameter, and the manufacturing cost is very high. Boron fiber is used almost exclusively in the United States in military aircraft and aerospace industries, and in Japan it is used only for sports and recreation equipment.

4.4.1.2   Carbon Fiber

Carbon fiber with a carbon content of more than 90% is obtained from sintering an organic fiber. Carbon fiber exhibits unique properties such as its light weight, high strength, good lubrication, and wear resistance. It costs approximately one-tenth the price of boron fibers. Preparation steps of carbon fiber include polyacrylonitrile (PAN) processing, thermosetting, carbonizing, graphitizing, etc. Commonly used carbon fibers include PAN fiber, asphalt fiber, and synthesized silk. PAN fiber is used in highly demanding aerospace applications. The less expansive asphalt fiber is finding application in automotive, machinery, electronics, and other areas.

4.4.1.3   Silicon Carbide Fiber

Silicon carbide fiber exhibits a high tensile strength and elastic modulus, good high-temperature strength and heat resistance, and excellent wetting between the metal and the fiber diameter. It is entirely possible to meet the requirements of 2,000°C heat resistance. There are mainly two methods to prepare silicon carbide fibers. Silicon carbide is deposited by CVD on the surface of a tungsten or carbon fiber; using silicone compound as the raw material, a continuous silicon carbide fiber can be obtained after heat treatment and sintering. Using ultrafine powder, binder, sintering additives, boron-doped silicon carbide fiber, and SiC (N) fiber can be prepared. Silicon carbide fiber can be used in heat-resistant materials and various types of reinforced fiber.

4.4.1.4   Alumina Fiber

Alumina fiber generally refers to continuous α-Al_2O_3 or γ-Al_2O_3 fiber containing Al_2O_3 as the main component and other materials such as SiO_2 and B_2O_3. Compared with carbon fiber, alumina fiber's mechanical strength is slightly lower, but it has excellent high-temperature mechanical properties and corrosion resistance, excellent electrical insulation, and high-temperature stability. Alumina fibers are prepared by a slurry method and a sol–gel method. In the slurry method, α-Al_2O_3 particles of less than 0. 5 μm are made into a liquid paste mixture with adhesive, and then the yarn is made by spinning the viscous liquid paste. Next, the yarn is subject to high-temperature sintering above 1,300°C, and the continuous alumina fiber is made. In the sol-gel method, alumina yarn is prepared by spinning viscous liquid made with alumina sol, silica sol, and boric acid, and the yarn is then sintered above 1,000°C, and the continuous γ-Al_2O_3 fiber is made. Alumina-fiber-reinforced aluminum-matrix composites are prepared in a pressure infiltration device composed of a vacuum chamber, furnace, pressure tank, graphite mold, and graphite piston. A preform made from loose alumina continuous fiber (as an enhanced agent) is put into a mold. Then the matrix alloy melt is infiltrated into preforms at a temperature of 1,073 K and a pressure of 14 MPa. The molded part is cooled from bottom to top along the direction of fiber solidification at a speed of 50 μm/s. The part is made into a 40-mm-long ingot at 2 to 3 Pa vacuum. The volume fraction of the fiber is about 50%. Using this pressure infiltration technique, Al-Si and Al-Cu composites are made. The anti-wear performance is analyzed by the solidification microstructure of the composite material. The results demonstrate that (1) the wear resistance of alumina-fiber-reinforced composites is enhanced by two to ten times compared with the unenforced material, but there are alumina fiber fractures and slips along the worn surfaces; (2) a hard silicon particle dispersion enhanced Al-27% (mass) Si hypereutectic alloy-matrix composite demonstrates higher abrasion resistance than an Al-7% (mass) Si alloy matrix because the hard phase may be connected with alumina fiber to prevent breaking of continuous alumina fiber due to surface wear; and (3) Al-Cu-based and Al-Si-based alloy-matrix failure hardening can effectively improve abrasion resistance. The hardness improvement of an α-phase matrix alloy can also improve abrasion resistance.

Cu/C_f Composite

Carbon fibers have a high longitudinal thermal conductivity (1,000 W/mK) and a small CTE (-1.6×10^{-6}/°C), and thus Cu/C_f packaging material has excellent thermal physical properties. Its thermal conductivity is influenced by the carbon-fiber content and is anisotropic. For composites with 40% carbon fiber content, the longitudinal thermal conductivity falls in a range of 230–220 W/mK, the transverse thermal conductivity in a range of 110–120 W/mK; and for composites with 60% carbon fiber content, the longitudinal thermal conductivity falls in a range of 160–150 W/mK, and the transverse thermal conductivity falls in the range of 40–45 W/mK. The vertical and horizontal thermal conductivities of Cu/C fiber packaging vary widely. On the other hand, the CTEs are also very different in the vertical and horizontal directions. For Cu/60% (volume fraction) carbon fibers, for example, the longitudinal CTE is 4.0×10^{-6}/°C, while the horizontal CTE is 15.5×10^{-6}/°C.

Liu et al. prepared Cu/C short-fiber packaging materials exhibiting isotropic physical properties by powder metallurgy. With carbon fiber content of 13.8%, 17.9%, 23.2% (volume fraction), the corresponding thermal conductivities (W/mK) achieved were 248.5, 193.2, and 157.4; the CTEs ($\times 10^{-6}/°C$) were 13.9, 12.0, and 10.8, respectively. In addition, using a porous 3D carbon fiber network preform and an argon-gas-assisted pressure infiltration method, isotropic Cu/C fiber packaging materials were prepared. For composites with Cu/72% (volume fraction) carbon, the CTE reached $(4–6.5) \times 10^{-6}/°C$ and the thermal conductivity reached more than 260 W/mK. Because the wettability of Cu and C is poor and the solid and liquid solubilities are small, they do not react and do not form carbides. Therefore, Cu/C interface connections are mainly based on mechanical bondings. Because there are no chemical reactions and no diffusion, the interfacial bonding is weak and the horizontal shear strength reaches only 30 MPa. Therefore, to obtain good Cu/C fiber packaging materials, the top priority is to solve the solubility problem between the two components. In addition, carbon fibers are expensive. For example, PITCH2120 carbon fiber has a thermal conductivity of two times that of Cu, but it costs approximately $2,000/kg. In addition, Cu/C fiber packaging materials have thermal expansion lagging issues.

Aluminum Graphite Materials

Conventional low-CTE materials like CuW, CuMo, AlSiC, Mo, and Kovar have reached physical and economic limitations and can no longer offer breakthroughs in cost and performance. Thus, innovative material solutions are sought to keep pace with emerging electronic cooling requirements. As a result, advancements in graphite fiber technology is spawning new ways to design and manufacture enhanced cooling solutions for electronics—especially airborne or handheld applications where low mass is critical.

A leader in developing aluminum graphite materials is Metal Matrix Cast Composites in Waltham, MA.

Aluminum and its alloys are extensively used in a large number of industrial applications due to their excellent combination of properties, e.g., high strength-to-weight ratio, good corrosion resistance, better thermal conductivity, and high deformability. Because of their high strength-to-weight ratio, automobile and aircraft components are generally manufactured out of aluminum alloys to make moving vehicles lighter, which results in savings in fuel consumption. However, the use of aluminum alloys as an antifriction material has been limited because of unfavorable wear. They tend to seize up when running under boundary lubrication conditions. To circumvent this limitation, i.e., to improve wear resistance, it has been proposed to disperse graphite particles in aluminum matrices. This will not only increase wear resistance, but it will also ameliorate the damping capacity and machinability of the base alloy.

Graphite is well known as a solid lubricant, and its presence in aluminum-alloy matrices makes the alloys self-lubricating. The reason for the excellent tribological properties of graphitic aluminum is that aluminum-alloy matrices yield at low stresses and deform extensively, which enhances the deformation and fragmentation of the surface and subsurface graphite particles even after a short running-in period. This provides a continuous film of graphite on the mating surfaces, which prevents metal-to-metal contact and, hence, seizing up. However, the basic problem associated with the production of aluminum-graphite composites is that the graphite particles are not wetted by the aluminum melt. Hence, for the successful entry of the graphite particles into the aluminum melt, either wettability should be induced or sufficient energy must be supplied to allow these particles to overcome the energy barrier at the gas–liquid interface.

Initial efforts confirmed that graphite particles could not be readily introduced into molten aluminum either by manual plunging or by injection below the bath surface. However, after a series of experiments and constant efforts, the conditions for wetting between graphite particles and aluminum melt have been evolved. The ultimate aim of the present investigation was to induce wetting between graphite particles and aluminum alloy melt using simple liquid metal technology and to develop potential components for automobile and engineering applications.

Dispersion of graphite particles in aluminum melt can be achieved only when the particles are wetted by molten aluminum. If the particles are not wetted, they remain floating on the top surface of the molten metal, maintaining a separate identity. Initial attempts at producing aluminum-graphite composites were restricted to the use of coated graphite particles either by nickel or by copper. Coating on graphite particles increases the surface energy and hence reduces the energy for complete immersion of a single graphite particle into the melt. This renders the process costlier and cumbersome and also limits the amount of heat. However, the process has successfully dispersed uncoated graphite particles in aluminum matrices. It has been up-scaled to the level of commercial heats, and castings of intricate shapes have been successfully made on a quality and quantity basis. Additionally, the inclined and off-center stirrer that was advocated in the initial experiments has been replaced by a vertical centrally located stirrer. This adds to the advantages of using a standard graphite crucible.

The aluminum-graphite composite melt has been successfully cast using shell molding, gravity, and pressure die casting techniques. In die casting, solidification is reasonably rapid and multidirectional, and there is limited time for undesirable floating of the graphite particles due to lower density as compared to the aluminum melt.

High-performance graphite fibers with axial conductivities of two to three times that of copper are being used to reinforce Al, Cu, and Mg alloys to enhance thermal conductivity and to restrict the CTE to specified values. In continuous spool form, high-modulus, pitch-based graphite fibers are expensive—on the order of $1,200/lb—and require weaving to produce a functional preform architecture that could be metal or resin infiltrated that would result in the desired properties. In woven form, the geometry and overall design capability of components is limited—mainly to niche-based, high-value-added applications.

The aluminum alloys typically used are clean modified in the crucible versions of A356 (7%Si and higher), depending on the application. In all alloy systems, the

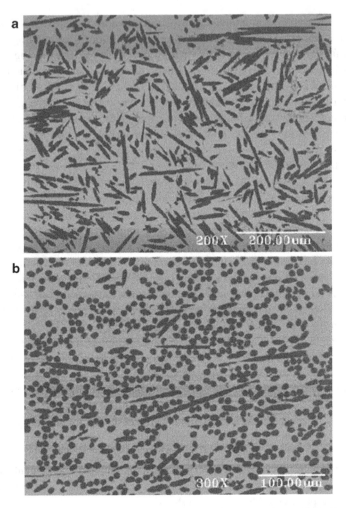

Fig. 4.10 SEM microstructure of milled graphite-fiber-reinforced aluminum (**a**) Section polished parallel to device mounting surface (in-plane or "x-y" section). (**b**) Section polished normal to device mounting surface (through-plane), or "z" section (Pictures courtesy Metal Matrix Cast Composites, used with permission)

highest conductivity values are measured from pure metals. Si is the principal precipitant in A356-type alloys and has a low solubility in Al at operating temperatures. Fortunately, Si has almost as high a conductivity as Al. Hence Al-Si alloys suffer less conductivity degradation from extensive alloying than any other system. Since Si inhibits the aluminum carbide formation by competing for carbon, the fiber-to-matrix bond is enhanced through Si to carbon bonding at the interface during contact with molten Al during the casting process (Figs. 4.10 and 4.11, Table 4.9).

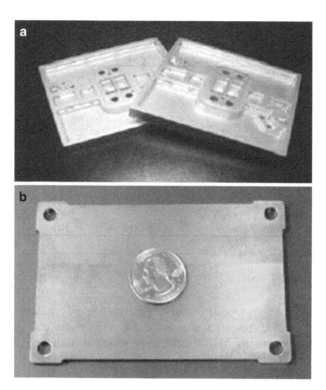

Fig. 4.11 Picture of some aluminum and copper graphite composite heat sinks. (**a**) Ka-band TR module housings for Boeing Spaceway phased-array antenna (photos courtesy Metal Matrix Cast Composites, used with permission). (**b**) Copper-graphite composite IGBT base plates

Table 4.9 Typical properties of aluminum-graphite composites

Composition	Al-70% graphite	Al-40% graphite
Thermal conductivity (W/mK)	200	230
CTE (ppm/K)	7.5	4
Density (g/cm³)	2.46	2.4

4.5   Sandwich Structure of Composite Materials

This type of stacked packaging material is generally divided into three layers. The middle layer is a low-expansion material, whereas the outside layers are materials with high electrical conductivity and thermal conductivity. Of course, there are two-layer, four-layer, and multilayer laminates. Generally, rolling and plating are used to prepare the sandwich structure. This type of composite has good thermal conductivity and low expansion coefficient. Basically there is no problem in densification.

Table 4.10 Typical CMC properties (Courtesy Torrey Hills Technologies)

Cu/Mo/Cu (ratio)	Density (g/cm³)	CTE(×10⁻⁶ K⁻¹)	Thermal Conductivity (W/mK)	
			In-plane	Through-plane
13:74:13	9.88	5.6	200	170
1:4:1	9.75	6.0	220	180
1:3:1	9.66	6.8	244	190
1:2:1	9.54	7.8	260	200
1:1:1	9.32	8.8	305	220

Table 4.11 Typical CPC properties (Courtesy Torrey Hills Technologies)

Cu/Mo70Cu/ Cu (ratio)	Density (g/cm³)	Young's modulus (GPa)	CTE (×10⁻⁶ K⁻¹)	Thermal conductivity (W/mK)	
				In-plane	Through-plane
1:4:1	9.5	195	7.0–9.0	280	180

In addition, the material processing cost is relatively low. For example, continuous production of rolled composite Cu/Invar/Cu composite plates can greatly reduce production costs. Thin foils 70 μm thick can be widely used in the PCB core layer and the lead frame material. At present, commercialized "sandwich" structure composite products include copper/molybdenum-copper/copper, Cu/Kovar/Cu, Cu/steel/Cu, and so on.

4.5.1 Copper/molybdenum/copper (CMC) and Copper/ molybdenum-copper/copper (CPC) Materials

Copper/molybdenum/copper (CMC) and copper/molybdenum-copper/copper (CPC) are laminates with a low-CTE material (Mo or Mo-Cu) sandwiched by thin copper sheets. The typical structure has three layers. According to some reports, some four or five layers were fabricated. Typical fabrication methods are hot rolling lamination, explosive lamination, and plating lamination. New developments on laser-assisted coating welding and ultrasonic welding have also been reported. The core material used in CPC is Mo70-Cu, which has a higher thermal conductivity than pure Mo. Thus the thermal conductivity of CPC composite materials is improved, especially in the out-of-plane direction.

The highlights of CMC and CPC are as follows (Tables 4.10 and 4.11):

(1) Very strong interface bonding, which can repeatedly resist 850 heat shock;
(2) Tailorable CTE matching that of semiconductor and ceramic materials;
(3) High thermal conductivity and no magnetism;
(4) Can be stamped into components.

Fig. 4.12 CMC and CPC microstructures

Table 4.12 Properties of CIC composite materials (Courtesy Torrey Hills Technologies)

Cu/Invar/ Cu (ratio)	Density (g/cm³)	Young's modulus (GPa)	CTE ($\times10^{-6}$ K^{-1})	Thermal conductivity (W/mK)	
				In-plane	Through-plane
1:1:1					
1:2:1					
1:3:1	8.5	–	7.0–9.0	110	48

The microstructure of CMC and CPC is shown in Fig. 4.12.

These types of laminates enjoy good thermal conductivity and low CTE. In addition, unlike W-Cu and Mo-Cu composites, there is no porosity issue. Relatively speaking, the manufacturing cost of CMC and CPC is lower than that of W-Cu and Mo-Cu. Furthermore, CMC and CPC laminates can be fabricated in large format (e.g., 24×24 in.).

From a metallurgical point of view, Cu and Mo are very different. Again, the melting point of Mo is higher than the boiling point of Cu. Cu has very low solubility in Mo. In addition, Mo and Cu sheet metals have different mechanical properties. The recrystallization temperature of Mo is >1,200°C, while copper recrystallizes at a temperature of <400°C. Therefore, it is difficult to laminate Cu and Mo. If not handled properly, the material tends to delaminate.

4.5.1.1   CIC(Cu/Invar/Cu) Thermal Management Materials

For Invar (Fe-Ni alloy) at temperatures below the Curie point, the "Invar" effect causes low thermal expansion. The "sandwich" structure made with Invar exhibits a low CTE. However, as the thermal conductivity of Invar is poor, the "sandwich" structure made with Invar exhibits low thermal conductivity in the out-of-plane direction. The properties of CIC are listed in Table 4.12.

References

1. Zweben C (2006) Thermal materials solve power electronics challenges. Power Electronics Technology, pp 40–47
2. http://www.advceramics.com/downloads/documents/85505.pdf
3. German RM, Hons KF, Johnson JL (1994) Powder metallurgy processing of thermal management materials for microelectronic applications. Int J Powder Metall 30(2):205–215
4. Zweben C (1992) Metal-matrix composites for electronic packaging. JOM 44(7):P15–P24
5. Zweben C (1998) Advances in composite materials for thermal management in electronic packaging. JOM 50(6):47–51
6. Yang F, Zhao YM (2001) State of research and development and trend of epoxy resins for electronic packaging. Electron Process Technol 22(6):238–241
7. Yu X, Wu R, Zhang G (1994) State of research and development and trend of metal matrix composites electronic packaging. Mater Rev 3:64–66
8. Xia Y, Song Y, Cui S, Lin C, Han S (2008) Preparation of Mo-Cu and W-Cu alloys and performance characteristics. Rare Met 32(2):240–244
9. Mu K, Kwong Y (2002) Mo-Cu material properties and applications. Metal Functional Mater 9(3):26–29
10. Zhang J, Hua X, Zhou X (2002) Electronic packaging SiC//Al composite prepared by infiltration and penetration mechanism. Metal Functional Mater 9(1):26–28
11. Ali ZA, Drury OB, Cunningham MF (2005) Fabrication of Mo/Cu multilayer and bilayer transition edge sensors. IEEE Trans Appl Superconduct 15(2):52–69
12. Guo-qin C, Long-tao J, Gao-hui W et al (2007) Fabrication and characterization of high density Mo/Cu composites for electronic packaging applications. T Nonferr Metal Soc China 17(1):580–583
13. Klein TW, Withers PJ (1996) Introduction to metal matrix composites. Metallurgical Industry Press, Beijing, pp 124–125
14. Johnson JL (1999) Powder metallurgy processing of Mo-Cu for thermal management applications. Int J Powder Metall 35(8):39–48
15. Hongwei X et al (2005) Interfacial reactions in 3D2SiC network reinforced Cu2 matrix composites prepared by squeeze casting. Mater Lett 59(12):1563
16. Lin Z et al (2008) Microstructure and thermomechanical properties of pressureless infiltrated SiCp/Cu composites. Compos Sci Technol 68(13):2731
17. Zhu J, Liu L, Hu G (2004) Composite electroforming of Cu//SiCp composite. Chinese J Nonferr Metal 14(1):84
18. Hyun-Ki K, Suk-Bong K (2006) Thermal decomposition of silicon carbide in a plasma-sprayed Cu/SiC composite deposit. Mat Sci Eng A 428(122):336
19. Martinez V et al (2003) Wetting of silicon carbide by copper alloys. J Mater Sci 38(19):4047
20. Yuang-Fa L, Sheng-Long L (1999) Effects of Al additive on the mechanical and physical properties of silicon reinforced copper matrix composites. Scr Mater 41(7):773
21. Levin L et al (1995) The mechanisms of phase transformation in diffusion couples of the Cu2Si system. Mater Chem Phys 40(1):56
22. Jun jun C, Yuepeng F, Mingbo T (2008) The latest progress of microelectronic packaging materials. Semicond Tech 33(3):185–189
23. Hodson TL (1995) AlN Steps up, takes the heat and delivers. Elect Packag Prod 35(7):26–30
24. Liang Q, Zhou HP, Fu RL et al (2003) Thermal conductivity of AlN ceramics sintered with CaF2 and YF3. Ceram Int 29:893–897
25. Larson SE, Slaby J (2004) Comparison of various substrate technologies under steady state and transient conditions. Integrated Electron Syst Sector 2:648–653

26. Vaed K, Florkey J, Akbar SA et al (2004) An additive micromolding approach for the development of micromachined ceramic substrates for RF applications. J Microelectron Mech Syst 30(13):514–521
27. Li XL, Ma HA, Zuo GH et al (2007) Low temperatures interring of high density aluminium nitride ceramics without additives at high pressure. Sci Mater 56(12):1015–1018
28. Zaosheng Z, Zhengya L, Zhiwu C (2008) Ceramic substrates for electronic packaging materials. Mater Guide 22(11):16–20

Chapter 5
Properties of WCu, MoCu, and Cu/MoCu/Cu High-performance Heat Sink Materials and Manufacturing Technologies

Abstract In this chapter, we will first present the properties of WCu, MoCu, and Cu/MoCu/Cu high-performance heat sink materials in great detail. Then we will introduce manufacturing technologies such as high-temperature liquid-phase sintering, reactive sintering, and infiltration. Then we will introduce CMC/CPC composite manufacturing technologies such as hot rolling lamination, explosive forming, and laser cladding. Finally, we will discuss the WCu and MoCu microelectronics packaging material manufacturing technologies research diagram and present detailed descriptions of each step.

Tungsten copper and molybdenum copper are not really alloys. Both of them are composed of two immiscible metallic phases. Therefore, they show characteristics of both metals, and their coefficients of thermal expansion (CTEs) and thermal conductivity (TCs) are complementary for each other, and they have good overall performance. The MoCu composite materials show combinational properties such as high electrical and thermal conductivities, low coefficient of thermal expansion, nonmagnetic and good high-temperature performance, and many others. They have a bright future in the field of electronic packaging applications. Compared with the traditional packaging materials, they have high thermal conductivity, and their CTEs closely match those of contacting materials. Their CTEs can be adjusted, which is a strong advantage; compared with SiC/Al composite materials, a variety of mechanical cutting processing can be performed with MoCu materials. Compared with W/Cu materials, molybdenum in MoCu has a relatively low density; thus the use of MoCu can reduce the weight of packaging materials. Because of this unique advantage, MoCu is a preferred choice in the aerospace, instrumentation, and portable equipment industries.

G. Jiang et al., *Advanced Thermal Management Materials*,
DOI 10.1007/978-1-4614-1963-1_5, © Springer Science+Business Media New York 2013

Table 5.1 Typical properties of heat sink grade WCu (Courtesy Torrey Hills Technologies)

Type	Composition			Properties
	Tungsten content (wt%)	Density (g/cm³)	CTE (ppm/K)	Thermal conductivity (W/mK)
W90Cu	90 ± 1	17	6.5	180–190
W85Cu	85 ± 1	16.3	7	190–200
W80Cu	80 ± 1	15.6	8.3	200–210
W75Cu	75 ± 1	14.9	9	220–230

Fig. 5.1 Some typical WCu heat sink parts (Courtesy Torrey Hills Technologies)

5.1   Properties of WCu, MoCu, and Cu/MoCu/Cu

5.1.1   W-Cu Thermal Management Material

Many grades of W-Cu materials are currently being used as heat sinks. By adjusting the content of tungsten, we can have its CTE designed to match those of materials such as ceramics (Al_2O_3, BeO), semiconductors (Si), and metals (Kovar). Their typical technical properties are shown in Table 5.1.

There are varieties of WCu products and specifications because there are varieties of electronic components and different component structures. Some typical WCu heat sink parts are shown in Fig. 5.1.

Their microstructures are shown in Fig. 5.2.

5.1.2   Copper Molybdenum (MoCu) Thermal Management Materials

As with W-Cu, the CTE of MoCu can also be tailored by adjusting the content of molybdenum. MoCu is much lighter than W-Cu, so it is suitable for aeronautic and astronautic applications (Table 5.2).

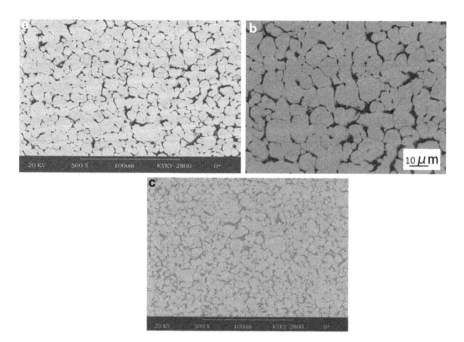

Fig. 5.2 (**a**) W-10Cu microstructure. (**b**) W-15Cu microstructure. (**c**) W-20Cu microstructure

Table 5.2 Typical properties of heat sink grade MoCu (Courtesy Torrey Hills Technologies)

Physical property	Mo50Cu50	Mo60Cu40	Mo70Cu30	Mo80Cu20	Mo85Cu15
Composition (wt%)	50% Mo Cu: balance	60% Mo Cu: balance	70% Mo Cu: balance	80% Mo Cu: balance	85% Mo Cu: balance
Density at 20°C (g/cm³)	9.5	9.6	9.7	9.9	10.0
CTE at 20°C (ppm/K)	9.9	9.5	7.5	7.2	6.8
Thermal conductivity (W/mK)	250	215	195	175	165
Specific heat at 100°C (J/kgK)	323	310	301	–	275
Specific electrical resistance 20°C (μΩm)	0.028	–	0.37	–	–
Young's modulus at 20°C (GPa)	172	–	225	21	248
Flexural strength (MPa)			–	1,103	1,138
Vicker's hardness (HV 10)	150	–	170	–	–

Table 5.3 Typical properties of heat sink grade CMC (Courtesy Torrey Hills Technologies)

Cu/Mo/Cu	Density (g/cm³)	CTE (ppm/K)	Thermal conductivity (W/mK)	
			In-plane	Through-thickness
13:74:13	9.88	5.6	200	170
1:4:1	9.75	6	220	180
1:3:1	9.66	6.8	244	190
1:2:1	9.54	7.8	260	210
1:1:1	9.32	8.8	305	250

Table 5.4 Typical properties of heat sink grade CPC (Courtesy Torrey Hills Technologies)

Cu/Mo70Cu/Cu	Density (g/cm³)	CTE (ppm/K)		Thermal conductivity (W/mK)	
		X-direction	Y-direction	In-plane	Through-thickness
1:4:1	9.46	7.2	9	340	300

5.1.3   CMC

Cu/Mo/Cu (CMC) is a sandwiched composite comprising a molybdenum core layer and two copper clad layers. It has an adjustable CTE, high TC, and high strength (Table 5.3).

5.1.4   CPC

Cu/Mo70Cu/Cu (CPC) is a sandwiched composite similar to Cu/Mo/Cu comprising a Mo70-Cu alloy core layer and two copper clad layers. It has different CTEs in the X and Y directions. Its thermal conductivity is higher than those of W/Cu, Mo/Cu, and Cu/Mo/Cu, and it is much cheaper (Table 5.4).

5.2   WCu Composite Manufacturing Technologies

The most common fabrication method is powder metallurgy (PM). More specifically, there are three different manufacturing techniques:

- High-temperature liquid-phase sintering,
- Reactive sintering,
- Infiltration.

5.2.1   High-temperature Liquid-phase Sintering

Due to the fact that the melting points of copper and tungsten are so different, it is possible to use high-temperature liquid-phase sintering to prepare the composite material. The advantage is its simple and mature process. The basic process steps include powder mixing, dry press, and sintering. The disadvantages are its high sintering temperature, long sintering cycle, and relatively low sintered body density (typically at 90 to 95% theoretical density). To obtain usable materials for heat sinks, high-temperature liquid-phase sintered W-Cu materials are often further processed using forging, hot pressing, and other methods. The additional postsintering process limits the use of high-temperature liquid-phase sintering in the fabrication of W-Cu heat sinks. Bhalla reportedly achieved good results using explosive compaction. It was also revealed that copper particle size played an important role in high-temperature liquid-phase sintering. The smaller the copper particle size, the higher the sintered density.

5.2.2 Reactive Sintering

Reactive sintering is commonly used to sinter tungsten powders. Similar methods are adopted to fabricate WCu composites. The typical sintering aids are Pd, Ni, Co, Fe, and others. The addition of sintering aids can reduce the sintering temperature and time and increase the sintered density significantly. Among the sintering aids for W-Cu composites, Co and Fe are the best. Using W90Cu (90% W) as an example, when Co content is 0.35% and the sintering is done at 1,300°C for 1 h, the resultant sintered composite has a 99% theoretical density, hardness of 300 HV, and flexural strength of 300 MPa. Ni and Pd are not as good as Co and Fe due to the fact both Ni and Pd can form alloys with Cu because Ni and Pd are infinitely soluble in molten copper, while Co and Fe are only partially soluble. During the sintering process, Co and Fe will form an intermetallic compound to promote the densification of tungsten. Unfortunately, the very addition of sintering aids reduces the electrical conductivity and TC significantly. They are rarely used in manufacturing W-Cu heat sink materials.

5.2.3 Infiltration

Infiltration starts with the preparation of a dry-pressed tungsten skeleton; the skeleton is then infiltrated with molten copper. The capillary force is employed to fill the micro-cavities inside the green body, provided that the molten metal wets the green body skeleton.

The advantages of infiltration are the high density and excellent electrical conductivity and TC that it produces in materials. The disadvantages are multiple machining steps after infiltration that can lead to higher cost and lower yield. Due to the good performance of the obtained materials, the infiltration method is the most common way to manufacture W-Cu heat sinks.

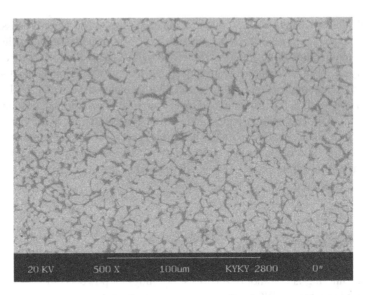

Fig. 5.3 Typical microstructure of WCu parts fabricated by dry press and infiltration

One of the key processes to make good W-Cu heat sink materials is to prepare an evenly distributed, good copper wetting, and high-purity tungsten skeleton. Tungsten powders are relatively hard to form, and a green body, microcracks, and delaminations are common challenges. There are three ways to prepare tungsten skeletons:

- High-temperature sintering
  The high-temperature sintering method involves the preparation of a low-density green body by dry pressing tungsten powders at a low pressure; then the green body is sintered at about 2,000°C on reducing the atmosphere to obtain a higher-density tungsten skeleton. This method entails a lower dry-press pressure, evenly interstice distributed skeleton, and good skeleton strength. During the sintering process, some of the micropores (about 6–8%) are closed and cannot be filtrated by molten copper. The density is about 92–94% of theoretic density.
- Dry press
  The dry press forming method is one of the most mature powder metallurgy methods. Many methods have been developed to improve the forming properties of tungsten powders, such as mixing copper powder, using copper-coated tungsten, and adding forming agents like rubber base or wax. Figure 5.3 shows a typical microstructure of W-Cu parts fabricated by dry press and infiltration.
- Mechanical alloying
  Mechanical alloying (MA) is a solid-state powder processing technique involving repeated cold welding, fracturing, and rewelding of powder particles in a high-energy ball mill. Originally developed to produce oxide-dispersion-strengthened (ODS) nickel- and iron-based superalloys for applications in the aerospace industry, MA has been shown to be capable of synthesizing a variety

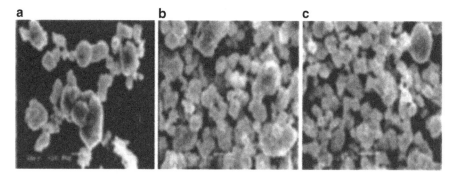

Fig. 5.4 The micrographs of mechanically alloyed Cu-W powders after ball milling at different times, (**a**) 10h (**b**) 40h (**c**) 60h

of equilibrium and nonequilibrium alloy phases starting from blended elemental or prealloyed powders. Typically MA is done in a reducing atmosphere. The drawback is the Fe impurities introduced during the ball mill stage, which affects the thermal and electrical performance of the resultant W-Cu material. Figure 5.4, 5.5, and 5.6 shows the micrographs of MA CuW powder, XRD, and W-Cu composite.

5.3   MoCu Composite Manufacturing Technologies

Molybdenum's melting point of 2,610°C is higher than copper's boiling point 2,595°C. Copper is not soluble in molybdenum. Most MoCu composites are fabricated using powder metallurgy, and the process is very similar to that of W-Cu. Generally, there are three manufacturing methods:

1. High-temperature liquid-phase sintering
   The basic high-temperature liquid-phase sintering process is very similar to that of W-Cu. The precursor materials can be either pure copper and molybdenum powders, or they can be a mixture of pure and oxide copper and molybdenum powders. The basic process flow is the same as that of W-Cu and basically consists of powder mixing, dry press, and sintering. This process works best for high-molybdenum-content composites since it is difficult to do by infiltration. For high-molybdenum-content composites, it is very common to use ultrafine molybdenum and copper powders as precursor materials and mechanical surface activation to promote sintering densification. The drawback is that molybdenum grains tend to have excessive growth. The structure of the resultant MoCu composite is not as homogeneous as that obtained by infiltration.
2. Infiltration
   As in the case of W-Cu, most MoCu heat sink materials are fabricated by infiltration. This process starts with dry pressing molybdenum powders to a

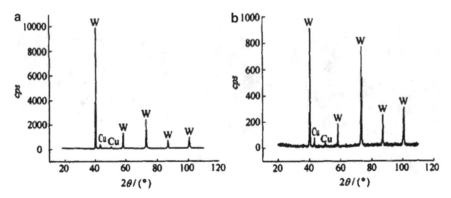

Fig. 5.5 The XRD graphs of mechanically alloyed Cu-W powders after ball milling at different times, (**a**) 10h (**b**) 40h

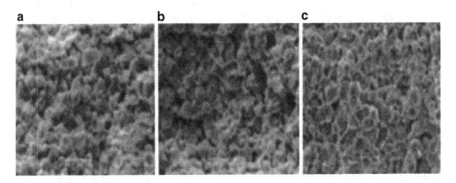

Fig. 5.6 The resultant WCu composite after mechanical alloying after ball milling at different times, (**a**) 10h (**b**) 40h (**c**) 60h

green body and presintering them in a reducing atmosphere to obtain a porous molybdenum body. Then molten copper infiltrates the micropores of the molybdenum skeleton to form a MoCu composite. The copper content is determined by the volume of interstices within the molybdenum skeleton. Since the copper density is very close to that of molybdenum, the volume of interstice is limited, and consequently the copper content is limited to about 30% or less by weight. The best way to fabricate lower-copper-content composites (like 85% Mo-15% Cu) is still by high-temperature liquid-phase sintering.

3. Mechanical alloying

MoCu composites fabricated by MA exhibits higher electrical conductivity, higher thermal conductivity, and higher hardness. During the MA process, metal particles undergo extensive stress, strain, and dislocation to form some nano-scale grain boundaries. The system has much higher residual energy, up to 10 KJ/ mol or more. After MA, the powders are highly activated. The longer the ball milling time, the smaller the metal particles and the larger the specific surface area. At the same time, there are many crystal surface and grain boundary defects. The powders are in a mesostate and very easily densified during sintering.

5.4 CMC/CPC Composite Manufacturing Technologies

The most commonly used manufacturing technologies include hot rolling lamination, explosive forming, and laser cladding.

5.4.1 Hot Rolling Lamination

Hot rolling lamination uses huge rolling machine pressure, aided by heat, to apply tremendous shear force to laminating materials. The surface oxide layers are broken down by the shear force, undergo plastic deformation, and form an atomic-level intermetallic bond. In the hot rolling lamination process, to make CMC and CPC, four key factors are important in controlling the obtained laminate quality, material surface condition before lamination, first rolling thickness deformation, rolling temperature, and annealing temperature.

Hot rolling lamination is a relatively mature manufacturing method with a low volume manufacturing cost. However, the initial capital investment is high.

Due to the fact that Cu and Mo are mutually insoluble, the bonding mechanism at the MoCu interface is a microscopic-level mechanical bonding.

5.4.2 Explosive Forming

Explosive forming is a metalworking technique in which an explosive charge is used instead of a punch or press. It can be used on materials for which a press setup would be prohibitively large or require an unreasonably high pressure. In addition, it is generally much cheaper than building a large enough and sufficiently high-pressure press; on the other hand, it is unavoidably a batch process, producing one product at a time and with a long setup time.

There are various approaches to explosive forming; one is to place a metal plate over a die, with the intervening space evacuated by a vacuum pump, place the whole assembly underwater, and detonate a charge at an appropriate height above the plate. For complicated shapes, a segmented die can be used to produce in a single operation a shape that would require many manufacturing steps or require that the material be manufactured in parts and welded together, with a concomitant loss of strength at the welds. There is often some degree of work hardening from the explosive-forming process, particularly in mild steel.

Tooling can be made out of fiberglass for short-run applications, out of concrete for large parts at medium pressures, or out of ductile iron for high-pressure work; ideally the tooling should have higher yield strength than the material that is being formed, which is a problem since the technique is usually only considered for material that is itself very hard to work.

Explosive forming was used in the 1960s for aerospace applications, such as the chine plates of the SR-71 reconnaissance plane and various Soviet rocket parts; it continued to be developed in Russia, and the organizing committees of such events as EPNM tend to contain many members from the former Soviet Union. It proved particularly useful for making high-strength corrugated parts that would otherwise have had to be milled out of ingots much larger than the finished product.

5.4.3   Laser Cladding

Laser cladding is a method of depositing material by which a powdered or wire feedstock material is melted and consolidated by use of a laser in order to coat part of a substrate or fabricate a near-net shape part.

The powder used in laser cladding is normally of a metallic nature and is injected into the system by either coaxial or lateral nozzles. The interaction of the metallic powder stream and the laser causes melting to occur and is known as the melt pool. This is deposited onto a substrate; moving the substrate allows the melt pool to solidify and thus produces a track of solid metal. This is the most common technique; however, some processes involve moving the laser/nozzle assembly over a stationary substrate to produce solidified tracks. The motion of the substrate is guided by a CAD system that interpolates solid objects into a set of tracks, thereby producing the desired part at the end of the trajectory.

5.4.4   WCu and MoCu Microelectronics Packaging Material Manufacturing Technologies Research Diagram

A research diagram of W-Cu and MoCu microelectronics packaging material manufacturing technologies is given in Fig. 5.7.

Descriptions of the preceding process steps are listed below.

1. W and Mo powder processing
   W and Mo powders have a large surface area and high activity. The surfaces will be oxide if they are exposed to the atmosphere. Because surface oxides cannot be wetted, they will adversely affect the quality of the following process steps. To remove the surface oxides, the powders should be heated at 900°C for 1 h.
2. Powder preforming
   Before performing, additives are often added to the powder mixtures to improve the pressing process. Additives are used to improve the mechanical strength of the

Fig. 5.7 Research diagram of W-Cu and MoCu microelectronics packaging material manufacturing technologies

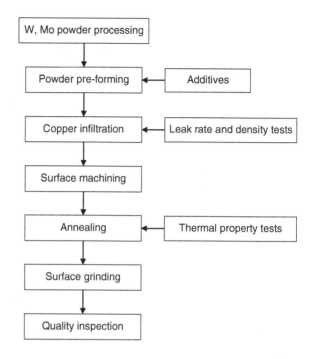

performs or to prevent segregation of the powders. They are removed before sintering or during sintering. Sometimes they are called binders. Additives can reduce friction between W powder particles and lower the preforming pressure. As a result, the performance capability is improved. There are many choices for additives, such as wax rubber, SBP, alcohol, acetone, stearic acid, oleic acid, and others.

Additives are utilized to improve the preform property, so high pressing pressure can be used.

3. Copper infiltration

Infiltration starts with the preparation of a dry-pressed tungsten skeleton; the skeleton is then infiltrated with molten copper. A capillary force is employed to fill the micro cavities inside the green body, provided that the molten metal wets the green body skeleton.

The infiltration temperature is 1,350°C and the duration is 1 h. After the infiltration process is completed, the leak rate, density, and CTE are tested.

4. Surface machining

After infiltration, there is extra copper on the surface of W(Mo)-C. The surface is normally milled. If the surface roughness, flatness, and parallelism are not up to customer requirements, the materials will be polished or grinded further. Finally, the work piece is machined according to customer design using CNC equipment.

5. Nickel plating

In the microelectronics field, nickel is widely used as a bottom barrier layer. A Ni layer can stop the interdiffusion of certain elements between the substrate and the coating. At the same time, the reliability of microelectronic components is improved by the Ni barrier. For W-Cu and MoCu, the Ni layer can improve the weldability between W-Cu/MoCu substrates and silver-copper solder. Ni can be deposited by plating or by an electroless method according to the customer's requirements. The thickness can be up to 100–300 mm. After the Ni is deposited and heated to 850°C for 10–15 min, there should be no air bubbles on the surface under a 40× amplification microscope.

6. Quality inspection

There should be no visible burrs, pinholes, or air bubbles on the surface under a 10× to 40× microscope. The surface scratch depth should be less than 12.5 μm, and the dent depth should be less than 30 μm.

References

1. Johnson JL, German R (1994) The solubility criterion for liquid phase sintering. Adv Powder Metall Particular Mater, 3:267–269
2. Johnson JL, German RM (1993) Phase equilibria effects on the enhanced liquid phase sintering of tungsten-cooper. Metall Trans A 24A:2369–2377
3. German RM (1993) A model for the thermal properties of liquid-phase sintered composites. Metall Trans A 24A:1745–1752
4. Hens KF, Johnson JL, Randall M (1994) German pilot production of advanced electronic packages via powder injection molding. Adv Powder Metall Part Mater 4:217–229
5. Teledyne Advanced Materials Nashville R&D Facility (1995) SEM evaluation of two selected Hosokawa test runs. Nashville, TN, 7 Feb 1995
6. German RM, Hens KF, Johnson JL (1994) Powder metallurgy processing of thermal management materials for microelectronic applications. Int J Powder Metall 30(2)
7. Petzow G, Kaysser WA, Amtenbrink M (1981) Liquid phase and activated sintering: theory and practice. In: Proceedings of the 5th international round table conference on sintering, Portoroz, Yugoslavia, 7–10 Sept 1981
8. Li CJ, German RM (1984) Enhanced sintering of tungsten-phase equilibria effects on properties. Int J Powder Metall Powder Technol 20(2), American Powder Metallurgy Institute
9. Li C, German RM (1983) The properties of tungsten processed chemically activated sintering. Metall Trans 14A:2031–2041
10. German RM, Munir ZA (1976) Enhanced low-temperature sintering of tungsten. Metall Trans 7A:1873–1877
11. German RM, Ham V (1976) The effect of nickel and palladium additions on the activated sintering of tungsten. Int J Powder Metall Powder Technol 12(2)
12. German RM, Munir ZA (1976) Systematic trends in the chemically activated sintering of tungsten. High Temp Science 8:267–280
13. Hayden HW, Brophy JH (1963) The activated sintering of tungsten with group VIII elements. J Electrochem Soc 110(7):805–810
14. Burton JJ, Machlin ES (1976) Prediction of segregation to alloy surfaces from bulk phase diagrams. Phys Rev Lett 37(21):1433–1436
15. Kaysser WA, Petzow G (1985) Present state of liquid phase sintering. Powder Metall 28(3):145–150
16. German RM, Munir ZA (1977) Rhenium activated sintering. J Less-Common Met 53:141

17. German RM, Munir ZA (1976) Temperature sensitivity in the chemically activated sintering of hafnium. J Less-Common Met 46:333–338
18. German RM, Munir ZA (1978) Heterodiffusion model for the activated sintering of molybdenum. J Less-Common Met 58:61–74
19. German RM (1983) A quantitative theory of diffusional activated sintering. Sci Sinter 15(1):27–42
20. Huppmann WJ, Riegger H (1975) Modelling of rearrangement processes in liquid phase sintering. ACTA Metall 23:965–971
21. Samsonov GV, Yakovlev VI (1967) Activated sintering of tungsten with palladium additions. Institute of Materials Science. Academy of Sciences of the Ukrainian SSR. Translated from Poroshkovaya Metallurgiya, 7(55), pp 45–49. Original article 30 Aug 1966
22. Samsonov GV, Yakovlev VI (1967) Activated sintering of tungsten with nickel additions. Institute of Materials Science, Academy of Sciences of the Ukrainian SSR. Translated from Poroshkovaya Metallurgiya, 8(56), pp10–16. Original article submitted 12 Apr 1966
23. Samsonov GV, Yakovlev VI (1969) Theory and technology of sintering, thermal, and chemicothermal treatment processes--activation of the sintering of tungsten by the iron-group metals. Institute of Materials Science, Academy of Sciences of the Ukrainian SSR. Translated from Poroshkovaya Metallurgiya, 10(82), pp 32–38. Original article submitted 21 May 1968
24. Samsonov GV, Yakovlev VI (1970) Activation of the sintering process of tungsten by the platinum-group metals. Institute of Materials Science, Academy of Sciences of Ukrainian SSR. Translated from Poroshkovaya Metallurgiya, 1(85), pp 37–44. Original article submitted 29 Jul 1968
25. Munir ZA, German RM (1977) A generalized model for the prediction of periodic trends in the activation of sintering of refractory metals. High Temp Sci 9:275–283
26. Gessinger GH, Fischmeister HF (1972) A modified model for the sintering of tungsten with nickel additions. J Less-Common Met 27(2):129–141
27. Kaysser WA, Zivkovic M, Petzow G (1985) Shape accommodation during grain growth in the presence of a liquid phase. J Mater Sci 20:578–584
28. German RM (1987) The two-dimensional connectivity of liquid phase sintered microstructures. Metall Trans A 18A:909–914
29. Yang S-C, Mani SS, German RM (1990) The effect of contiguity on growth kinetics in liquid-phase sintering. JOM 42(4):16–19
30. Brophy JH, Hayden HW, Wulff J (1961) The sintering and strength of coated and co-reduced nickel tungsten powder. Trans Metall Soc AIME 221(6):1225–1231
31. German RM (1996) Microstructure of the gravitationally settle region in a liquid-phase sintered dilute tungsten heavy alloy. Metall Mater Trans A 26A:279–288
32. Bhalla AK, Williams JD (1976) A comparative assessment of explosive and other methods of compaction in the production of tungsten-copper composites. Powder Metall 1:31–37
33. Moon IH, Lee JS (1977) Sintering of W-Cu contact materials with Ni and Co dopants. Powder Metall Int 9(1):23–24
34. Huppmann WJ, Bauer W (1975) Characterization of the degree of mixing in liquid-phase sintering experiments. Powder Metall 18(36):249–258
35. Parikh NM, Humenik M Jr (1957) Cermets: II, wettability and microstructure studies in liquid-phase sintering. J Am Ceram Soc 40(9):315–320
36. Kothari NC (1967) Densification and grain growth during liquid-phase sintering of tungsten nickel-copper alloys. J Less-Common Met 13:457–468
37. Wittenauer J, and Nieh TG (1991) Fine-grained W-Cu-Co alloys via liquid phase sintering. Lockheed Missiles & Space Co. In: Crowson A, Chen ES (eds) Tungsten and tungsten alloys-recent advances. The Minerals, Metals & Materials Society, Warrendale, PA
38. Teodorovich OK, Levchenko GV (1964) Nickel in tungsten-copper contacts. Institute of Materials Problems, Academy of Sciences, Ukrainian SSR. Translated from Poroshkovaya Metallurgiya, 6(24), pp 43–47. Original article submitted 28 Jan1964

39. Naidich YV, Lavrinenko IA, Evdokimov VA (1977) Liquid phase sintering under pressure of tungsten-nickel-copper composites. Institute of Materials Science, Academy of Sciences of the Ukrainian SSR. Translated from Poroshkovaya Metallurgiya, 4(172), pp 43–49. Original article submitted 14 Jul 1976
40. Stevens AJ (1974) Powder-metallurgy solutions to electrical contact problems. Powder Metall 17(34):331–346
41. Gessinger GH, Melton KN (1977) Burn-off behaviour of W-Cu contact materials in an electric arc. Powder Metall Int 9(2):67–72
42. Srikanth V, Upadhyaya GS (1983) Effect of tungsten particle size on sintered properties of heavy alloys. . Indian Institute of Technology, Kanpur (India). Received 19 Aug 1983; Revised 26 Oct 1983
43. Zukas EG, Rogers PSZ, Rogers RS (1976) Spheroid growth by coalescence during liquid-phase sintering. Z Metallkde, 67:591–595
44. Grinberg EM, Tikhonova IV, Ol'shanskii BI, Ol'shanskii AB, Zapol MY (1986) Reaction of carbon with molybdenum during indirect sintering. Tulachermet Scientific-Production Association. Tula Polytechnic Institute. Translated from Poroshkovaya Metallurgiya, 8(284), pp 20–25. Original article submitted 19 Nov 1985
45. Prokushev NK, Smirnov VP (1986) Kinetics of densification and growth of refractory phase grains in the liquid-phase sintering of very finely divided tungsten-copper materials. Institute of Materials Science, Academy of Sciences of the Ukrainian SSR. Translated from Poroshkovaya Metallurgiya, 9(285), pp 30–37. Original article submitted 28 Jan 1986
46. Buchatskii LM, Stolin AM, and Khudyaev SI (1986) Kinetics of the change of density distribution in hot one-sided pressing of a viscous porous body. . Department of the Institute of Chemical Physics, Academy of Sciences of the USSR, Chernogolovka. Translated from Poroshkovaya Metallurgiya, 9(285), pp 37–42. Original article submitted 28 Jan 1986
47. Skorokhod VV, Panichkina VV, Prokushev NK (1986) Theory and technology of sintering, thermal, and chemicothermal treatment processes. Structural inhomogeneity and localization of densification in the liquid-phase sintering of tungsten-copper powder mixtures. Institute of Materials Science, Academy of Sciences of the Ukrainian SSR. Translated from Poroshkovaya Metallurgiya, 8(284), pp 14–19. Original article submitted 13 Nov 1985
48. Skorokhod VV, Solonin YM, Filippov NI, Roshchin AN (1983)Theory and technology of sintering, thermal, and chemicothermal treatment processes: sintering of tungsten-copper composites of various origins. . Institute of Materials Science, Academy of Sciences of Ukrainian SSR. Translated from Poroshkovaya Metallurgiya, 9(249), pp 9–13. Original article submitted 30 June 1982
49. Panichkina VV, Sirotyuk MM, Skorokhod VV (1982). Theory and technology of sintering, thermal, and chemicothermal treatment processes: liquid-phase sintering of very fine tungsten-copper powder mixtures. Institute of Materials Science, Academy of Sciences of the Ukrainian SSR. Translated from Poroshkovaya Metallurgiya, 6(234), pp 27–31. Original article submitted 31 Jul 1981
50. Sebastian KV, Tendolkar GS (1979) High density tungsten-copper liquid phase sintered composites from coreduced oxide powders. Int J Powder Metall Powder Technol 15(1):45–53
51. Moon IH, Lee JS (1979) Activated sintering of tungsten-copper contact materials. Powder Metall 22(1):5–7
52. Kothari NC (1982) Factors affecting tungsten-copper and tungsten-silver electrical contact materials. Powder Metall Int 14(3):139–159
53. Johnson JL, German RM (1994) Chemically activated liquid phase sintering of tungsten-copper. Int J Powder Metall 30(1):91–102
54. Huppmann WJ (1975) Sintering in the presence of liquid phase. In: Proceedings of the forth international conference on sintering and related phenomena, University of Notre Dame, Notre Dame, IN, 26–27 May 1975
55. German RM, Rabin BH (1985) Enhanced sintering through second phase additions. Powder Metall 28:7–12

56. Kosco, New low-expansion alloys for semiconductor applications. Solid State Technology. Jan 1969, 47-49
57. Terasawa M, Minami S, Rubin J, Kyocera Corporation (1983) A comparison of thin flim, thick film, and co-fired high density ceramic multilayer with combined technology: T&T HDCM (thin film and thick film high density ceramic module). Int J Hybrid Microelectron 6(1)
58. Zweben C (2006) Thermal materials solve power electronics challenges. Power Electronics Technol, pp 40–47
59. Yang X, Yueqing S (2008) Preparation and properties of Mo-Cu and W-Cu alloys. Chin J Rare Met 32(2):240–244
60. Han L, Daocheng L, Zhengyun W et al (2007) Influence of high power ball mill technology on structure of tungsten-copper composites. Powder Metall Ind 17(2):30–33
61. Degan X, Hui C, Xicong L (2006) Advances in research on aluminum silicon carbide electronic packaging composites and components. Mater Rev 20(3):111–115

Chapter 6
Novel Methods for Manufacturing of W85-Cu Heat Sinks for Electronic Packaging Applications

Abstract High-velocity compaction (HVC) is a production technique that uses a high-speed punch motion on powder materials to achieve superior mechanical properties. However, as the particle size of the powder metals decreases, achieving higher-density numbers becomes increasingly challenging. For this reason, the HVC technique cannot be adopted for manufacturing W85-Cu heat sinks as the particle size of the tungsten powder is very fine (FSSS 3 μm or finer). In this study, the HVC process at elevated temperatures is studied for W85-Cu heat sinks (HTHVC). Tungsten skeletons prepared by conventional uniaxial methods are compacted by the HTHVC method at various elevated temperatures. The compacted tungsten skeletal blanks are infiltrated with copper at an elevated temperature of 1,350°C for 2 h. The mechanical properties including density, thermal conductivity, hermeticity, and coefficient of thermal expansion, have been found to be in line with the requirements for W85-Cu heat sinks.

6.1 Introduction

To obtain W-Cu electronic packaging materials with better performance, it is desirable to use smaller particle size tungsten powder, expecting a tungsten skeleton to achieve 72–80% relative density, while right now only 55–60% relative density is possible using conventional pressing techniques. If the compressive pressure is further increased, blanks will have cracks or undergo delamination and the desired shape cannot be produced. A new forming technology or process is required to improve the density of tungsten powder skeletons.

Back in the 1980s, Chinese Engineering Academy member Huang Peiyun put forward the concept of high-speed compaction techniques [1], and the mechanisms of high-velocity compaction were discussed. He believes that the reason that high-velocity compaction can result in high density are twofold. First, when pressed by dynamic pressure instead of static pressure, the powder body is impacted by not only the static pressure but also momentum; the greater the velocity, the greater the momentum. As the impact time is very short and the impact pressure on the powder is much greater than the static force, densification efficiency is high. Second, the powder deforms fast in the HVC process, generally faster than the work hardening rate, and the deformation rate is not hampered by the work hardening effect. Therefore, the pressure needed for densification in the HVC process is less than the level needed during the static densification process.

High-velocity compaction (HVC) is a technology launched by a Swedish company, Höganäs, in June 2001. In this technology, a hydraulic hammer is employed to produce a strong shock wave in a very short period of time (20 ms or so) and the powder is pressed into high density. It is a breakthrough for the powder metallurgy industry seeking low-cost, high-density material preparation technologies. This technology is still under development, even though it is used for pressing copper powder, iron powder [2–5], stainless steel powder [6], polymers [7, 8], and other conventional powders; there are no reports used for refractory metal powders such as tungsten powder or molybdenum powder.

Fine tungsten powder of FISS 2.8 μm particle size is used as the raw material. The tungsten skeleton is prepared by a high-temperature, high-velocity compaction (HTHVC) technique. In this newly developed technology, the powder is first pressed into a low-density green body in a static press. The low-density green body is then presintered for a certain amount of time. The presintered skeleton is then put in a mold for high-velocity treatment. At this point, the density of the black piece is further increased. Because the heating temperature is usually selected below the recrystallization temperature of the powder for high-velocity treatment, this technology is termed HTHVC.

The main purpose in this study is to solve the problem of not being able to meet the tungsten green body relative density requirement of 72–80% at room temperature. This density value of 72–80% is required for the packaging industry. The temperature can vary from 600°C to 950°C. As the melting point of tungsten is high, 950°C is still below the recrystallization temperature of tungsten powder.

Copper-tungsten composites have excellent thermal conductivities and match the coefficient of thermal expansion (CTE) of Al_2O_3, BeO, and other ceramics commonly used in electronics packaging. Among different copper-tungsten materials, W85-Cu (85% W) is widely used in the packaging of microwave, RF, and optical communication power amplifiers [9–14]. One of the primary requirements for W85-Cu heat sinks is to have higher densities and higher airtightness or hermeticity [15–18]. To achieve higher density it is very important that the particle size of the tungsten remain very small. The smaller particle size of tungsten also helps in achieving a homogeneous WCu composite material with improved Cu and W distribution. However, due to their poor plasticity, even in the presence of a molding agent, manufacturing W85Cu composites from fine W particles using conventional dry press or isostatic press methods becomes challenging. The obtained green density is normally 60% of the theoretical density, making it unsuitable for W85-Cu heat sink applications, which require about 72% of theoretical density. On the other hand, when the pressure is increased, the green body tends to delaminate and the obtained green density continues to remain at 65%.

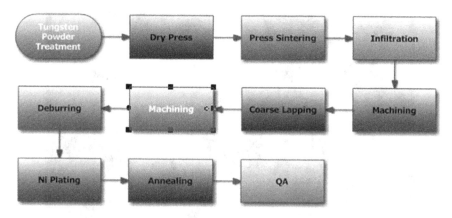

Fig. 6.1 Flow chart of traditional method of manufacturing WCu composites

The traditional method of manufacturing WCu involves preparing a tungsten skeleton from fairly larger tungsten particles using a dry press method. The skeleton is sintered and later infiltrated with copper to form a copper-tungsten composite. The process is described elaborately in Fig. 6.1. Maintaining a uniform material distribution with larger tungsten particles is very challenging. For this reason, this study has focused on using novel manufacturing methods that could use smaller tungsten particles and still meet the required density number.

In HVC, a high-speed punch motion is used to achieve higher densities, typically using the same powder materials. The HVC method has the capacity to significantly improve the mechanical properties of powder metallurgy (PM) parts. Several investigations indicate that high-density components can be obtained using HVC. Also, low ejection force and uniform density are obtained using HVC. Powder filling, compaction, and ejection of green compacts are similar to uniaxle compaction presses. The application of HVC to tungsten powder metallurgy has not been reported due to the poor plasticity of compacted tungsten powders.

In this research, the combined merits of high-speed compaction [19–21] and high-temperature pressure forming have been used to devise a new forming method, HTHVC. In the current method, the interaction between high-temperature and high-velocity compaction has been studied in detail. The new method has been named HTHVC since its temperature is higher than tungsten's ductile-brittle transition temperature (200–400°C) while still below the recrystallization temperature.

6.2 Experiments

Using a proprietary mold-forming agent, SBP, and the conventional static dry press method, FSSS 2.83 μm tungsten powers (Fig. 6.2) were pressed into blanks measuring 51×23×15 mm. The resultant relative density was found to be 61% that of pure tungsten. The green body was later heated at 900°C in a reducing atmosphere to

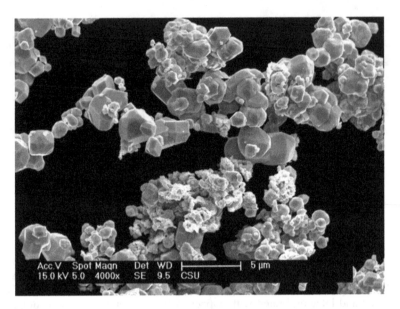

Fig. 6.2 SEM pictures of Fisher FSSS 2.83 μm tungsten powers

burn out the binder matter. After burning the binder out, the blank was heated to 650°C and 950°C in a reducing atmosphere. A 60-ton of high-speed friction press was then used to further increase the density of the tungsten skeleton. The tungsten skeleton was further infiltrated with molten copper at 1,350°C for 2 h.

The prepared specimen per the preceding specification were machined into parts in accordance with the standards of electronic packaging materials. The prepared samples were later characterized using the following instruments: JR-3 laser flash to measure thermal conductivity and thermal diffusivity, HELIOT3068 type He adsorption for hermeticity measurement, DHL-402PC to measure the CTE, and SIR200 SEM to evaluate the electron microscopy microstructure.

6.3   Experimental Results

6.3.1   A. Effects of Temperature on HTHVC W Skeleton Density

To study the effect of temperature on the HTHVC process, tungsten skeletons with a starting green compact density of 11.65 g/cm³ were subjected to an HTHVC process at temperatures of 25°C, 600°C, and 950°C, respectively. Table 6.1 shows the effect of temperature on the obtained W skeleton green density.

The results of this experiment illustrate clearly that temperature has a significant effect on the density of the 85Wcu composite. At elevated temperatures, the final density of the green skeleton was found to surpass 14 g/cm³. This is in line with the density requirement of 13.98 g/cm³ for electronic packaging applications. However,

Table 6.1 Effect of temperature on obtained tungsten skeleton green density

Temperature (°C)	Green blank density (g/cm³)	Density after HTHVC processing (g/cm³)	Improvement (%)
25	11.65	12.96	11.24
600	11.65	14.15	21.45
950	11.65	14.39	23.51

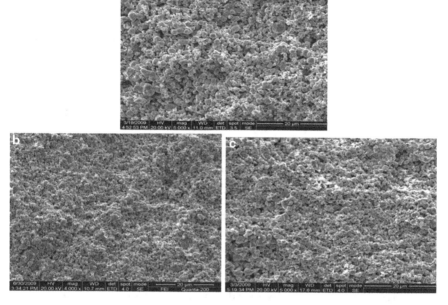

Fig. 6.3 SEM pictures of fractured surfaces of green blanks after HTHVC processing. (**a**) 25°C. (**b**) 600°C. (**c**) 950°C

the same process at room temperature (HVC) failed to meet the density requirement. The density increase at elevated temperatures can be attributed to the increase in plasticity of the tungsten particles at higher temperatures. Thus the HTHVC process helps in fulfilling density requirements even with finer tungsten particles versus larger tungsten particle sizes (7–8 μm) and higher tonnage press used in the conventional process. Figure 6.3 shows SEM pictures of fractured surfaces of green blanks after HVC processing at different temperatures.

6.3.2 B. Effects of Precursor Tungsten Skeleton Density on HTHVC W Skeleton Density

From the results of the foregoing experiment, the temperature was set at 950° C for further experiments on the HTHVC process. In the next round of investigation, the

Table 6.2 Effects of precursor tungsten skeleton density on HTHVC W skeleton density

	Group 1	Group 2	Group 3	Group 4
Precursor tungsten skeleton density (g/cm³)	10.59	11.59	11.65	13.26
After HTHVC process (g/cm³)	13.85	14.28	14.39	14.62
Improvement (%)	30.78	23.20	23.51	10.25

Table 6.3 Effect of precursor skeleton density on final density

	Precursor tungsten skeleton		After HTHVC processing	
Item	Density (g/cm³)	Relative density (% pure W)	Density (g/cm³)	Relative density (% pure W)
1	11.77	60.98	13.95	72.28
2	11.75	60.98	13.94	72.23
3	11.64	60.98	13.82	71.61
4	11.75	60.98	13.91	72.21
5	11.73	60.98	13.92	72.22

effect of the starting green density of the tungsten skeleton on the final density of the 85WCu composite was studied. For this, a series of samples with densities ranging from 10.59 to 13.26 g/cm³ were subjected to HTHVC at 950°C. After HTHVC processing, the final composite densities were measured and tabulated (Table 6.2).

From the preceding table it can be observed that HTHVC processing is very effective in compacting tungsten precursor skeletons. The increase slowed down as the precursor density increased. Table 6.2 data were the average of five samples. HTHVC also showed very good reproducibility (Table 6.3). The precursor skeleton blanks had an average density of 11.728 g/cm³ and a standard deviation of 0.051 g. After HTHVC processing, the average density was 13.908 g/cm³ and the standard deviation was 0.052 g.

6.3.3   C. HTHVC W85-Cu Microstructures and Physical Properties

After 950°C HTHVC processing, the resultant W skeleton was infiltrated with Cu in 1,350° for approximately 2 h. The microstructure of the composite can be observed in Fig. 6.3, which that the phase distribution of copper is uniform across the connected tungsten skeleton. Also, no obvious clustering of tungsten and copper was observed from the SEM images shown in Fig. 6.4.

The mechanical properties of the specimen are tabulated in Table 6.4. It can be observed that the W85-Cu samples processed using the HTHVC method have a relative density of 99.5%, TC coefficient of 185 W/mK, and hermeticity of $1 \times 10\text{-}10$ Pa • m³/s. Also, the CTE and other properties were found to be in conformance with the standards of heat sinks for electronic packaging applications.

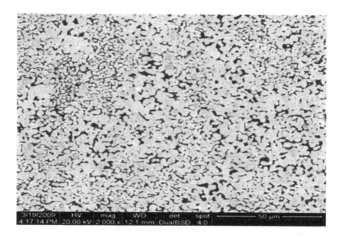

Fig 6.4 SEM micrographs of HTHVC-processed W85-Cu

Table 6.4 Physical properties of W85-Cu prepared by 950°C HTHVC

Relative density (%)	CTE ($\times 10^{-6}$/K)			Hermeticity(Pa·m³/s)	Thermal conductivity (W/mK)
	100°C	200°C	300°C		
99.5	6.94	7.60	7.71	1×10^{-10}	185

6.4 Discussion

6.4.1 HTHVC Densification

Tungsten powders are hard and brittle. It is difficult to form them using dry press. In the regular HVC process, due to the relatively large impact in the tungsten blank axis, there is a high-volume change and certain amount of axial elastic aftereffects. Per the theory of compaction, in the initial stage, the densification is caused by filling the porosity with powders (or removal of porosity). After that, the main densification mechanism is due to the deformation of powder. The densification mechanism of iron has been reported in the literature [22]. During HVC processing of iron particles, a significant amount of heat is produced due to friction between the iron particles. This results in higher local temperatures, which causes local neck-to-neck sintering between iron particles. This mechanism helps to increase the density with the HVC process for iron.

However, in the case of tungsten powder, because the ductile-brittle transition temperature of tungsten is very high, the melting point is also much higher than room temperature. At room temperature, W particle deformation is still dominated by brittle fracture. Hence the density gain by HVC over the conventional uniaxial dry press is very limited and the blank is prone to delamination. Thus processing tungsten at room temperature using HVC does not help in increasing the density significantly. However, at elevated temperatures, the tungtsten skeleton becomes

Fig. 6.5 High-amplification SEM micrographs of W skeleton prepared by 950°C HTHVC showing joined W particles

more ductile and the local contact temperature between particles also increases. At a forming temperature of approximately 950°C, which is higher than tungsten's ductile-brittle transition temperature, more tungsten particles would experience plastic deformation. At the same time, tungsten particles would thereby be partially sintered at the neck portion during the HVC process, increasing the compact density significantly. Figure 6.5 shows an SEM image of a tungsten skeleton formed using the HVC process at 950°C.

6.5   Advantage of HTHVC

In this work, blanks were first obtained using a static dry press and later heated in a reducing atmosphere. The temperature was selected such that it was higher than the plastic-brittle transition temperature yet lower than the recrystallization temperature of tungsten. Following an increase in temperature, the blanks were subjected to HVC. The resultant W skeleton after processing in HVC at the elevated temperature displayed significantly higher density, eliminating the need for special forming and lubricating agents and other powder-heating devices.

This process enjoys all the advantages of a typical HVC process including higher impact energy, high plasticity of powders, and higher density. The high-speed suppression helps in increasing the green density of the fine tungsten powder through increased impact energy and ensures stability of the powder bulk density. Therefore, the proposed high-speed, high-temperature, and high-pressure molding technology has all the advantages to be considered for manufacturing CuW heat sinks for electronic packaging applications.

6.6 Disadvantage of HTHVC

For a typical microelectronics packaging CuW heat sink part, here is the rough cost breakdown:

1. Material cost: Cu and W, 10–35% of total cost
2. Infiltration and machining: 10–35% of total cost
3. Finishing cost: Ni/Au plating, 30–50% of total cost

With the addition of the HTHVC process, the infiltration and machining cost portion is estimated to increase by 50–100%.

6.7 Conclusions

Based on the foregoing experimental results, the following conclusions were drawn:

1. At elevated temperatures, the HVC process increases the density of tungsten skeletons significantly.
2. Processing tungsten at elevated temperatures resulted in a relative density of 99.5%, airtightness of $1 \times 10\text{-}10$ Pa • m^3/s, and TC of 185 W/mK. Thus the mechanical properties were found to be in line with the requirement for heat sinks for electronic packaging applications.
3. The distribution of copper in the tungsten skeleton was uniform and no significant clustering was observed in the SEM images.
4. Thus, the HTHVC method for processing W85Cu composites is very useful in obtaining high-density composites with uniform material distribution and desirable mechanical properties.
5. The HTHVC process could potentially add significant cost to the CuW.

References

1. Huang P (1982) Principles of powder metallurgy. Metallurgical Industry Press, Beijing, pp 169–275
2. Chi Y, Guo S, Meng F, Xia Y, Heng Z, Dong L (2005) High-velocity pressing compaction technology of powder metallurgy. Powder Metall Ind 15(6):41–45
3. Barendvanden B, Christer F, Tomas L (2006) Industrial implementation of high velocity compaction for improved properties. Powder Metall 49(2):107–109
4. Jonsén P, Häggbladh HÅ, Troive L, Furuberg J, Allroth S, Skoglund P (2007) Green body behavior of high velocity pressed metal powder. Mater Sci Forum 534/536:289–292
5. Aslund C (2004) High velocity compaction (HVC) of stainless steelgas atomized powder[C]. In: Herbert D, Raimund R (eds) Euro PM 2004 conference proceedings, EPMA, Shrewsbury, pp 533–564
6. Bruska A, Bengt S, Leif K (2005) Development of a high-velocity compaction process for polymer powders. Polym Test 24(4):909–919

7. Jauffrès D, Lame O, Vigier G, Doré F (2007) Microstructural origin of physical and mechanical properties of ultra high molecular weight polyethylene processed by high velocity compaction. Polymer 48(21):6374–6383
8. Chen Z (2007) Modern powder metallurgy technology. Chemical Industry Press, Beijing, pp 326–328
9. German RM, Hons KF, Johnson JL (1994) Powder metallurgy processing of thermal management materials for microelectronic applications. Proc Int J Powder Metall 30(2):205–215
10. Zweben C (1998) Advances in composite materials for thermal management in electronic packaging. JOM 50(6):47–51
11. Crum S (1996) MCM substrate choices expand. Proc Electr Pack Prod 36(4):47–49
12. Qiang Z, Dongli S (2000) Recent achievements in research for electronic packaging substrate materials. Proc Mater Sci Technol 8(4):66–69
13. Zhengchun L, Zhifa W, Guosheng J (2001) Advances in metal-matrix material for electronic packaging. Proc Ordnance Mat Sci Eng 02:70–73
14. Xuebing Y, Renjie W, Guoding Z (1994) Research and development on metal-matrix electronic packaging material. Proc Mater Rev 3:64–66
15. Zhifa W, Guosheng J, Zhengchun L (1998) Ultrapressure forming and low-temperature sintering of tungsten. Proc Rare Metal Mater Eng 27(5):290–293
16. Guosheng J, Zhifa W, Hong W (2007) Study on control of pores in W-skeleton in preparing W-15Cu composite. Powder Metall Technol 25(2):126–128
17. Meifen L, Chaofei Z, Shukui L (2005) Effect of compressive deformation on microstructures and properties of infiltrated W-Cu composites. Proc Ordnance Mater Sci Eng 03:17–19
18. Zhengyun W, Daocheng L, Wei F (2007) Research of preparing W-Cu composites by high-energy ball milling. Proc Cemented Carbide 24(3):148–151
19. Skoglund P (2001) High density PM parts by high velocity compaction. Powder Metall 44(3):20–25
20. Richard F (2002) HVC punches PM to new mass production limits. Proc MPR 57(9):26–30
21. Hoganaos CE (2001) Promotes potential of high velocity compaction. Proc MPR 56(9):6–10
22. Zhou M, Rosakis AJ, Ravichandran G (1996) Dynamically propagating shear bands in impact loaded pre notched plates, Proc of experimental investigations of temperature signatures and propagation speed. Mech Phys Solid 44(6):981–1006

Chapter 7
Improved Manufacturing Process of Cu/Mo70-Cu/Cu Composite Heat Sinks for Electronic Packaging Applications

Abstract Looking back at the development of various thermal management materials, one can see the limit imposed by nature or by physical laws. Single-element materials, like Al and Cu, have their advantages and disadvantages. Composite materials like CuW, CuMo, AlSiC, and Al-graphite are a tradeoff between the thermal conductivity (TC) and coefficient of thermal expansion (CTE). Using CuW as an example, a higher Cu content will increase the TC and lower the weight (both are desirable). On the flip side, a higher Cu content will also lead to a higher CTE (undesirable). CuW has many compositions like 50/50, 60/40, 70/30, 80/20, 85/15, 87/13, 89/11, and 90/10. The various compositions reflect the compromise between the conflicting needs for CTE and TC. It was almost impossible to improve both CTE and TC until a Cu/Mo70Cu/Cu laminate material was developed. Cu/Mo70Cu/Cu is a laminate material with Mo70Cu sandwiched by two thin layers of Cu. During the rolling lamination process, Mo particles inside the Mo70Cu composite is elongated in the X-direction (the rolling direction). The resultant Cu/Mo70Cu/Cu is anisotropic. The X-direction has a lower CTE (like 7.0 ppm/°C for a 1:4:1 ratio), and the Y-direction has a higher CTE (~9.0 ppm/°C for a 1:4:1 ratio). Material physics still works here, coupled with relatively high TC (up to 300 W/mK for through thickness), but this unique material feature makes it very attractive for long and thin dies like LDMOS. This chapter provides a detailed discussion of the background and fabrication of this magical material.

7.1 Introduction

Copper-molybdenum/tungsten composites have excellent thermal conductivity, and their coefficients of thermal expansion (CTEs) match those of Al_2O_3, BeO, and other ceramics commonly used in electronics packaging. They are widely used in the packaging of laser diodes, microprocessors, microwaves, RF, and optical communication power amplifiers [1–6]. One of the primary requirements for heat sinks is that they must have higher relative densities and higher airtightness or hermeticity

G. Jiang et al., *Advanced Thermal Management Materials*,
DOI 10.1007/978-1-4614-1963-1_7, © Springer Science+Business Media New York 2013

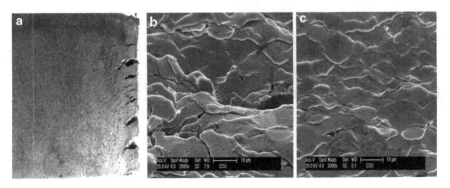

Fig. 7.1 (**a**) Mo70-Cu edge breakup during rolling process. (**b**) Mo70-Cu interfacial cracks between Mo and Cu particles after 800°C annealing. (**c**) Mo70-Cu interfacial cracks between Mo and Cu particles after 900°C annealing

[7–12]. Using the high-velocity compaction (HVC) process developed by Jiang and Kuang [13], the material properties of Mo70-Cu reach a density of 9.7 g/cm³, a thermal conductivity (TC) of 180 W/mK, and a CTE of 7.5×10^{-6} ppm/K. However, for some critical applications that require much larger TC, the materials' properties need further improvement.

In theory, the TC of CPC composite structures made from Mo70-Cu could be improved significantly. The manufacturing technology of core materials is described by Jiang and Kuang [13]. The CPC composite structure is hot rolled together. However, the manufacturing process of the core material is rather complicated. One of the primary challenges in making Cu/Mo70-Cu/Cu is the macroscale cracking of core material edges (Fig. 7.1a) during the rolling process. Edge breakup drastically increases manufacturing costs. The other serious problem in this manufacturing process is the presence of interfacial micro cracks between Mo and Cu particles after the core composite is hot rolled, which results in poor hermeticity and plasticity.

Copper and molybdenum are very different materials in their crystal structures, melting points, and CTEs. Below the melting point of copper, there is almost no mutual solubility; above the melting point of copper, a small amount of molybdenum can be dissolved in copper. Therefore, the Mo70-Cu alloy is actually a mechanical mixture of two metals. The plasticity of the alloy depends on the volume fraction of copper. It is generally believed that molybdenum-copper and tungsten-copper alloys with more than 30 wt% copper could be cold rolled and hot rolled into plates [14–16].

But if there are microvoids, small pores, or other defects, the defects soon become a high stress center and sources of cracking. The hermeticity and plasticity of the composite could decrease significantly. Macrocracks will appear on work piece surfaces.

Further, under elevated temperatures, Mo/Cu mechanical bonding strength in Mo70-Cu becomes weaker. At the surfaces and the edges of a work piece, the temperature is relatively lower compared to that at the center of the work piece. Since the CTE of Cu is high and it is about three times that of Mo, thermal expansion of the weakly linked Mo/Cu composite will be sensitive to temperature differences. In other words, thermal stress at the surfaces and edges of the work piece aggregates the crack propagation of the microcracks at the Cu-Mo interfaces due to plastic deformation and the microcracks due to microvoids and other defects. That should explain the severe edge cracking taking place at the surfaces and edges of the work piece. It should be pointed out that stress concentration at the edges also plays some role in edge cracking.

Another source of cracking at Mo and Cu interfaces comes from a work-hardening effect during the hot rolling processing of the Mo70-Cu core material. During the hot rolling process, the materials undergo significant plastic deformation. Dislocations in metal particles take place. As the grains are stretched and twisted, slid bands and twins are formed. As the grains are stretched further, densities of dislocations, slide bands, and twins increase; therefore, work hardening increases. Such changes in microstructures result in increased residual stress, deterioration in plasticity, and delinking of mechanical bonding between Mo and Cu particles. Therefore, molybdenum and copper interfaces are separated and become sources of cracks [17, 18], resulting in Mo70-Cu core material fracture.

Therefore, to reduce the internal residual stress and to eliminate the interfacial microcracks in materials, annealing is required. When the Mo70-Cu core is annealed at 800°C and 900°C, there is still a significant amount of microcracks. SEM images of Mo70-Cu interfacial microcracks between Mo and Cu particles after 800°C and 900°C annealing are shown in Fig. 7.1b, c.

Here we will report results from optimizing the infiltration time and annealing temperature during the hot rolling process.

7.2 Experiments

1. Optimizing infiltration time.

   Infiltration time has been observed to be an important parameter governing the quality of the Cu/Mo70-Cu/Cu core. In this experiment, a 99.9% Mo material with a particle size of 4.8 μm and 99% Cu sheet were used for the composite. The traditional method of manufacturing Mo/Cu involves preparing a Mo skeleton from Mo particles using a dry press method. The skeleton is sintered and later infiltrated with copper to form a copper-molybdenum composite. While the infiltration temperature was set to 1,350°C, different infiltration times have been studied to understand its influence on the core. Two different infiltration times, 2 and 6 h, were examined in this experiment. The tensile tests of Mo70-Cu core samples were conducted using the INSTRON MODEL 8032 electronic universal testing machine. The samples were kept at 20°C, 150°C, 250°C, 350°C, 450°C and 550°C for 10–15 min. The elongation speed was 1 mm/min. A SEM image was obtained using a Sirion 200 field emission high-resolution scanning electron microscope.

2. Optimizing intermediate annealing during rolling process

Prior to rolling, the Mo70-Cu core is annealed at an elevated temperature. In this experiment, the influence of the intermediate annealing process on the quality of the CPC composite was studied. The effect of various annealing temperatures including 800°C, 900°C, 1,000°C, and 1,100°C were studied as a part of this experiment. The annealing time was set to 1 h for all the above-mentioned temperatures.

7.3   Results and Discussions

1. *Effect of Infiltration Time*

The densification effect of sintering temperature from 1,200°C to 1,500°C and sintering time from 1 to 4 h on the final density of the Mo-18Cu composite was investigated by Johnson [19]. It was found that there is significant copper evaporation once the temperature exceeds 1,400°C. The effect of infiltration time is discussed in this section. The effects of infiltration time on the edge surface characteristic are listed in Table 7.1.

It can be observed that the samples that underwent 2-h infiltration display poor plasticity, whereas samples that underwent 6-h infiltration exhibit excellent plasticity. The reason behind the poor plasticity for samples with 2 h infiltration time can be construed as weak infiltration of copper particles in the Mo skeleton. For samples that undergo 2-h infiltration, Mo skeletons were not fully infiltrated by Cu particles. This created numerous pores and weak bonding between the Mo and Cu particles, leading to poor plasticity. Even if there were only 10–30% deformation, serious edge cracking would take place (Fig. 7.2a).

On the other hand, for samples that undergo 6-h infiltration, the Cu particles are fully infiltrated into a Mo skeleton gap. This reduces the porosity defects within the core material. Also, the interfacial bonding strengths between molybdenum-molybdenum particles and molybdenum and copper particles are improved significantly. In particular, the Cu melt infiltrated into the Mo skeleton forms a uniform distribution of the networklike structure. Because the molybdenum

Table 7.1 Effects of infiltration time on Mo70Cu plasticity (the deformation rate is the ratio of the thickness difference before and after deformation to the thickness before the deformation, in percentage; the work pieces are hot rolled at 350°C)

Infiltration temperature	% Deformation	Edge surface characteristic
1,350°C infiltration for 2 h	10	Severe edge breakup
	20	Very severe edge breakup
	30	Surface crack
1,350°C infiltration for 6 h	10	Smooth surface
	20	Smooth surface
	30	Smooth surface
	40	Smooth surface
	50	Some edge cracks

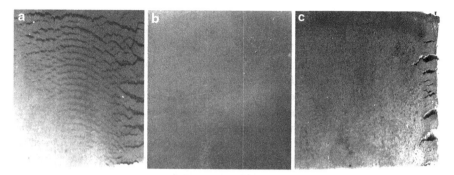

Fig. 7.2 Effect of infiltration on plastic deformation of core material. (**a**) 1,350°C, 2 h, 30% deformation. (**b**) 1,400°C, 6 h, 40% deformation. (**c**) 1,400°C, 6 h, 50% deformation

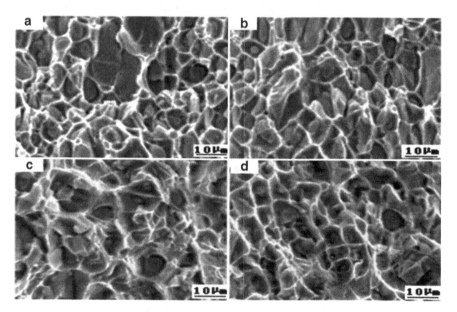

Fig. 7.3 SEM images of Mo70-Cu fracture surfaces at different temperatures: (**a**) 20°C. (**b**) 150°C. (**c**) 250°C. (**d**) 350°C

particles could move along with the copper particles during deformation, processing performance has been significantly improved. Even if the sample deformation after rolling increased to 40%, the surface shown in Fig. 7.2b would be in good condition. The edge crack shown in Fig. 7.2c does not appear until the deformation reaches 50%.

To gain insight into the mechanical bonding strength between Mo particles and Cu particles, tensile tests were performed at different temperatures. The tests show that Mo70-Cu core material has a tensile strength of 608 MPa and an elongation rate of 24.2% at room temperature, and the SEM images shown in Fig. 7.3a represent a ductile fracture surface. Cu fracture dimples and intergranular and transgranular fractures are clearly seen. At room temperature, copper has a tensile strength of about 400 MPa, and at room temperature Mo annealed at 1,350°C has a tensile strength of about 500 MPa [20]. It is inferred that the netlinking strength between Cu and Mo particles is strong.

As temperature is increased to 150°C, 250°C, 350°C, 450°C, and 550°C, the tensile strength of core materials decreases to 300, 250, 275, 200, 75 MPa, respectively, and the elongation rate is reduced from 22% to 14% almost linearly. The other SEM images of fracture surfaces are shown in Fig. 7.3b–d. At approximately 350°C, the tensile strength and elongation rates are relatively stable, and it is suitable for rolling deformation. From the fracture surfaces it can be seen that there are increased dimple depths, more intergranular fractures, and fewer transgranular fractures as the temperature is increased. Clearly, the netlinking strength of Cu and Mo particles is weakened as the temperature is increased.

2. *Effect of intermediate annealing during rolling process*

The effect of annealing temperature on the TC of W-Cu15 composites was investigated [8, 21]. It was found that the TC of W-Cu15 composites could be improved by annealing at 800°C. The SEM images of the Mo70-Cu microstructure after intermediate annealing at different temperatures during a rolling process are shown in Fig. 7.4. From the SEM images microvoids on both the sample annealed at 800°C and on the sample annealed at 900°C can be observed. Almost no microvoids are found on samples annealed at 1,000°C and 1,100°C. This could be explained in two ways. On the one hand, when Mo70-Cu is annealed at 800°C and 900°C, even though the Cu phase is sufficiently annealed, the Mo phase is insufficiently annealed. This is because pure Mo has a Brinell microhardness of 290 and 280 respectively at 800°C and 900°C and a Brinell microhardness of 205 at 1,100°C, according to Zhou [22]. On the other hand, copper has a melting temperature of 1,083°C. When the temperature reaches 1,100°C, the copper phase melts. At this temperature, its viscosity is still low enough to allow copper melt to move and fill the microcrack due to surface tension; its viscosity is still high enough not to leak out. After the core material is held at 1,100°C for 1 h, the bonding between the Mo phase and Cu phase is repaired and its plasticity is improved significantly. When the temperature is over 1,100°C, the copper melt could flow out of the Mo skeleton. This could leave the Mo70-Cu core material with small pores, which could result in a reduction in the density, the TC, and the electrical conductance.

3. *Composite Properties with Optimized Process Parameters*

The composite Cu/Mo70-Cu/Cu material is joined together in a hot rolling process using optimized process parameters. The SEM images of a cross section of Cu/Mo70-Cu/Cu are shown in Fig. 7.5. The joining mechanisms between copper and Mo70-Cu include mechanical bonding created by interlocking of rough surfaces and metallic bonding activated by heat energy released by severe plastic deformation during the hot rolling process [23–25]. From the low-amplification

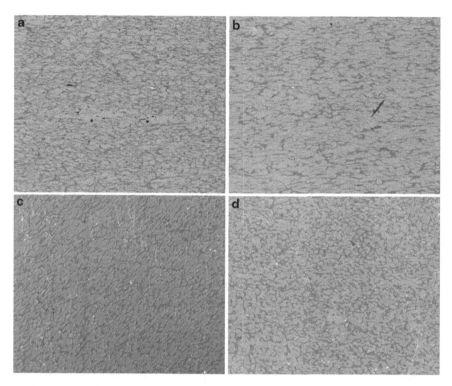

Fig. 7.4 SEM images (500×) of M70-Cu composite materials at four different annealing temperatures: (**a**) 800°C. (**b**) 900°C. (**c**) 1,000°C. (**d**) 1,100°C

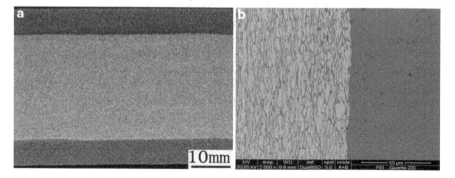

Fig. 7.5 SEM image of cross section of Cu/Mo70-Cu/Cu at (**a**) low magnification and (**b**) high magnification

graph shown in Fig. 7.5a, it is noticed that the interface is smooth and the waviness is small. The binding is tight and rigid without cracks and voids. The high-amplification graph in Fig. 7.5b shows the grain of Mo particles being stretched in the direction of the interface, which indicates severe plastic deformation.

Table 7.2 Physical properties of CPC composite materials

Cu/Mo70Cu/Cu	Density	TCE	TC	Hermeticity
Thickness ratio	g/cm³	×10⁻⁶/K	W/(m.K)	×10⁻¹⁰ Pa.m³/s
1:4:1	9.52	7.0–9.5	265	8.85

The through-plane TC of Cu/Mo70-Cu/Cu is measured by a JR-2 Laser Thermal Conductivity Analyzer. It is found that a sample with a 1:4:1 thickness ratio has a value of 265 W/mK. The theoretical calculation of the TC is included here to compare the calculated result to the measured result. We assume the TC of Cu, λ_1 is 400 W/mK and the TC of Mo, λ_2 is 140 W/mK [26]. For the Mo70-Cu core, the volume fraction of Cu, ϕ_1 is 32.8% and the volume fraction of Mo, ϕ_2 is 67.2%.

The calculated TC of Mo70-Cu is

$$\lambda_{core} = \phi_1\lambda_1 + \phi_2\lambda_2 = 225\left(\frac{W}{m.K}\right).$$

If we assume the total temperature difference across the entire thickness of CPC is the sum of all individual temperature differences across each layer and let $H_1 = 1/6$ and $H_2 = 4/6$, the calculated TC of Cu/Mo70-Cu/Cu could be estimated as

$$\lambda_{cpc} = \frac{\lambda_1\lambda_{core}}{2H_1\lambda_{core} + H_2\lambda_1} = 263.5\left(\frac{W}{m.K}\right).$$

The theoretical calculation is slightly smaller compared to the measured TC of a CPC composite. The mismatch suggests that the values of Mo and Cu TC used for this calculation are slightly higher than the actual values. If we use the measured TC value of 180 W/mK, then the calculated TC of the CPC would be 220 W/mK. This value is about 17% smaller than the measured results. This indicates that the TC of Mo70-Cu is improved after the core composites are hot rolled and annealed properly. Therefore, the improved manufacturing process could have a significant impact on the TC values of the core and the CPC. The other properties of CPC composite materials are listed in Table 7.2. The density is measured by water displacement. The CTE is measured using a Rigaku Thermal Expansion Analyzer. The leak test is conducted on a Nissan HELIOT 306S Helium adsorption test machine.

7.4   Conclusion

In this chapter, an improved manufacturing process of CPC composite heat sinks for electronic packaging applications has been presented. The properties of the composite material are improved by optimizing the infiltration time and annealing temperatures. The TC of the composite material is improved significantly compared to that of the core Mo70-Cu material. Also, the experimental result matches well the theoretical calculation. This composite material can be successfully used as a heat sink in electronics packaging industry.

References

1. German RM, Hons KF, Johnson JL (1994) Powder metallurgy processing of thermal management materials for microelectronic applications. Proc Int J Powder Metall 30(2):205–215
2. Zweben C (1998) Advances in composite materials for thermal management in electronic packaging. JOM 50(6):47–51
3. Crum S (1996) MCM substrate choices expand. Proc Electr Pack Prod 36(4):47–49
4. Zhang Q, Sun D (2000) Recent achievements in research for electronic packaging substrate materials. Proc Mater Sci Technol 8(4):66–69
5. Liu Z, Wang Z, Jiang G (2001) Advances in metal-matrix material for electronic packaging. Proc Ordnance Mater Sci Eng. (2):70–73
6. Yu X, Wu R, Zhang G (1994) The research and development on metal-matrix electronic packaging material. Proc Mater Rev 3:64–66
7. Wang Z, Jiang G, Liu Z (1998) Ultrapressure forming and low-temperature sintering of tungsten. Proc Rare Metal Mater Eng 27(5):290–293
8. Jiang G, Wang Z, Wu H (2007) Study on control of pores in W-skeleton in preparing W-15Cu composite. Powder Metall Technol 25(2):126–128
9. Liu M, Zhang C, Li S (2005) Effect of compressive deformation on microstructures and properties of infiltrated W-Cu composites. Proc Ordnance Mater Sci Eng (3):17–19
10. Wang Z, Luan D, Feng W (2007) Research of preparing W-Cu composites by high-energy ball milling. Proc Cemented Carbide 24(3):148–151
11. Skoglund P (2001) High density PM parts by high velocity compaction. Powder Metallurgy 44(3):20–25
12. Chen G, Zhu D, Zhan R, Zhang Q, Wu G (2005) Highly dense Mo/Cu composite by squeeze casting and their thermal conducting properties. Chinese J Nonferr Metals. 15(11):1864–1868
13. Jiang G, Wang Z, Gu Y, Zhang Q, Kuang K Novel methods for manufacturing of W85-Cu heat sinks for electronic packaging applications, submitted to IEEE Transactions on Components, Packaging and Manufacturing Technology 2(6):1039–1042
14. Cai Y, Liu B (1999) Densification of W-Cu composite powder metallurgy technology issues and approaches. Powder Metall Tech 17(2):138–144
15. Lu D (2004) W-Cu material production. Appl Develop China Tungsten Ind 19(5):69–74
16. Zhu W, Lu D (2005) W-Cu material present development status of the application and the production. Powder Metall Mater Sci Eng 10(1):21–25
17. Yang ZY, Wang F-C, Li S (1998) 93 W alloy under static tensile load and fracture characteristics of the gap effect. Journal of Materials Science and Engineering 21(5):2
18. Fan ZK, Liang S, Xu X (2001) Bond strength of W-Cu/CuCr integrated material. Trans Nonferr Metals Soc China 11(6):835–837
19. Jonson JL, German RM (2001) Role of solid-state skeletal sintering during processing of Mo-Cu composites. Metall Mater Trans A 32A:605
20. Jiang G (2010) Preparation processing optimizations of W-Cu and Cu/Mo70-Cu/Cu electronic packaging materials and its relative basic researches. Ph.D. Dissertation, Central South University, pp 73–75
21. He P, Wang Z, Jiang G, Cui D (2004–03) Effect of infiltrating temperature and annealing temperature on thermal conductivity of W-Cu composites. Mining Metall Eng 24(3):76–77
22. Zhou J (2008) CBC and CMC metal layered composite material, master's thesis, Central South University, pp 60–70
23. Zhu A, Wang K, Zhang B (2006) Electronic packaging with Cu/Mo/Cu composites technology. Rare Metals Letters 25(7):35–39
24. Li X, Qi K, Zhu Q (1999) Zinc/aluminum composite of rolling. Chinese J Nonferr Metals 9(2):300–304
25. Zhang S, Guo Z (1995) Al/Cu interfacial bonding composite plate rolling mechanism. Central South Univ Technol 26(4):509–513
26. Chen G, Wu G, Zhu D, Zhang Q (2005) The thermo-physical properties of high dense Mo/Cu composites fabricated by squeeze casting technology. In: 6th international conference on electronic packaging technology, Shenzhen, IEEE, 2005, pp 321–324

Chapter 8
AlSiC Thermal Management Materials

Abstract Aluminum silicon carbide (AlSiC) is a metal-matrix composite consisting of an aluminum matrix with silicon carbide particles. It has a relatively high thermal conductivity (166–255 W/mK), and its thermal expansion can be adjusted to match that of other materials, e.g., silicon and gallium arsenide chips and various ceramics. AlSiC composites are suitable replacements for copper-molybdenum (CuMo) and copper-tungsten (CuW) composite materials as heat sinks, where the application requires lower weight. AlSiC has about one-third the weight of copper, one-fourth that of CuMo, and one-sixth that of CuW. AlSiC is also stronger and stiffer than copper. Currently it is used as a heat sink for power electronics (e.g., IGBTs), heat spreaders, housings for electronics, and lids for chips, e.g., microprocessors and ASICs. This chapter serves as an introduction to AlSiC materials, covers its basic fabrication methods, and reviews its applications in microelectronics packaging.

8.1 Introduction

Aluminum is one of the most widely used materials in various industries. Its density is approximately one-third that of CuMo composites and approximately one-fifth that of WCu composites. Its low density and low cost make it one of the most widely used materials, especially for aerospace and auto applications. Its thermal conductivity (TC) is higher than that of most refractory metals such as tungsten, titanium, chromium, and molybdenum. Its TC is higher than those of WCu and CuMo composites and is about half that of copper, which has highest thermal conductivity among the most commonly used metallic materials. However, its mechanical properties are relatively low, and its coefficient of thermal expansion (CTE) is highest among the most commonly used metallic packaging materials. For microelectronics and optoeletronics packaging applications that require low weight, good TC, aluminum is the top choice. A comparison of aluminum properties compared to those of other metallic materials is given in Table 8.1.

G. Jiang et al., *Advanced Thermal Management Materials*, 109
DOI 10.1007/978-1-4614-1963-1_8, © Springer Science+Business Media New York 2013

Table 8.1 Metallic material candidates for packaging applications

Materials	CTE ($\times 10^{-6}$/°C)	TC (W/m.K)	Density ($\times 10^3$ Kg/m^3)	Yield stress (MPa)	Ultimate stress (MPa)
Al	23	237	2.71	20	70
Cu	16.6–17.6	410	8.94	55–330	230–380
W	4.3	174	19.3		1,400–4,000
Ti	8.6	21.9	4.54		500
Mo	5.4	140	10.2		
Ni	13.0	90.7	8.89	140–620	310–760
Cr	4.9	93.9	7.19		
Au	14	317	19.32		

Table 8.2 Process-mature ceramic material candidates for packaging applications

Material	Density (g/cm³)	CTE ppm/°C (25–150°C)	TC (W/mK)	Bend strength (MPa)	Young's modulus (GPa)
SiC	3.2	2.7	200–270	450	415
AlN	3.3	4.5	170–200	300	310
Alumina	3.98	6.5	20–30	300	350
Beryllia	3.9	7.6	250	250	345

Table 8.3 Properties of AlSiC composite material [2]

Material	Density (g/cm³)	CTE ppm/ (25–150°C)	TC (W/mK)	Bend strength (MPa)	Young's modulus (GPa)
AlSiC (63v% SiC)	3.0	7.5	170–200	450	175

Naturally, a composite composed of aluminum and enforcement materials with high mechanical strength and low CTE is desirable. There are many choices for enhancement materials. A comparison of SiC properties compared to those of other process-mature packaging materials is given in Table 8.2. SiC has the lowest density and lowest CTE. Its TC, bending strength, and Young's modulus are the highest in this group of materials. Furthermore, SiC is a top choice material for high-temperature applications. A device made from SiC can tolerate temperatures up to 700°C. Composite packaging materials made of Al and SiC can have low density, high mechanical strength, good TC, and low cost. Its CTE can be adjusted to match those of silicon and other process-mature packaging materials such as AlN, alumina, and beryllia.

In this chapter, we will cover aluminum matrix composites (AMCs), specifically on Al. Particle-reinforced aluminum matrix composites (PAMCs) have attracted interest in various fields including aerospace, automotive, sports, and microelectronic and optoelectronic packaging due to their high performance and comparatively low cost.

Aluminum silicon carbide (AlSiC) is gaining wide acceptance as an electronic packaging material mainly due to the fact that its CTE can be tailored to match that of Si or GaAs by varying the Al:SiC ratio while maintaining the TC and high strength-to-weight ratio. Silicon-carbide-particle-reinforced aluminum, commonly called AlSiC in the packaging industry, was used in microelectronic and optoelectronic packaging by C. Zweben and his colleagues at GE in the early 1980s [1].

An example of AlSiC composite materials [2] made by Ceramics Process Systems is given in Table 8.3. Its density is approximately one-third that of CuMo composites and approximately one-fifth that of WCu composites. Its TC is close to that of WCu and CuMo composite materials. Its CTE is close to those of silicon and other process-mature packaging materials.

In summary, the major advantages of AMCs compared to unreinforced materials are as follows:

- Enhanced mechanical strength
- Improved stiffness

- Reduced density (weight)
- Adjustable CTE
- Good TC
- Advanced thermal management properties

8.2   Types of Aluminum Matrix Composites

AMCs can be classified into four types depending on the type of reinforcement [3].

(a) Particle-reinforced AMCs
(b) Whisker-or short fiber-reinforced AMCs
(c) Continuous fiber-reinforced AMCs
(d) Monofilament-reinforced AMCs

8.3   Challenges and Solutions in Fabrication of AlSiC Thermal Management Materials

8.3.1   Manufacture of High SiC Volume Fraction AlSiC Composites by Pressure Infiltration of Liquid Aluminum into Packed SiC Particulate with a Bimodal Size Distribution

One of the most promising metal matrix composites (MMCs) for electronic packaging applications is a silicon-carbide-particle-reinforced AMC with a high volume concentration because the strength and stiffness of the composites generally increase with increasing reinforcement volume fraction. A more practical way to attain high particle volume fractions is to pack particles having different sizes. To achieve large particle volume fractions, particles of substantially different sizes have to be used. Molina [6] found that high volume fraction composites could be produced by infiltrating liquid aluminum into preforms made by mixing and packing SiC particles with average diameters of 170 and 16 μm. The maximum particle volume fraction (0.74) was attained for a mixture having 67% coarse particles (Fig. 8.1).

Figure 8.1 shows the particle volume fractions attained in this work versus the content of coarse particles in a mixture. The results indicate that particle volume fractions above 0.6 can be easily obtained. As expected, the compactness shows a maximum as the relative amount of the two particles is changed. A mixture having 67% coarse particles shows the maximum compactness, 0.74, a value 30% higher than those obtained for compacts containing only fine or coarse particles (0.55 and 0.58, respectively).

Fig. 8.1 Experimental results (*filled circles*) for particle volume fraction (*Vp*) of SiC compacts versus percentage of coarse particles in mixture. The *continuous line* denotes the fine-particle fraction and the *broken line* denotes the coarse-particle fraction

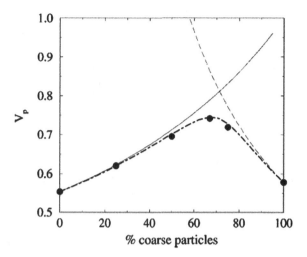

8.3.2 Processing of Advanced AlSiC Particulate Metal Matrix Composites Under Intensive Shearing: A Novel Rheo Process

One of the major challenges when processing MMCs is achieving a homogeneous distribution of reinforcement in the matrix as it has a strong impact on the properties and the quality of the material. To obtain a specific mechanical/physical property, ideally the MMC should consist of fine particles distributed uniformly in a ductile matrix and with clean interfaces between particle and matrix. However, current processing methods often produce agglomerated particles in the ductile matrix and as a result they exhibit extremely low ductility [7, 8]. Clustering leads to a nonhomogeneous response and lower macroscopic mechanical properties. Particle clusters act as crack or decohesion nucleation sites at stresses lower than the matrix yield strength, causing the MMC to fail at unpredictably low stress levels [9, 10]. Possible factors resulting in particle clustering are chemical binding, surface energy reduction, and particle segregation [11]. The current processing methods often produce agglomerated particles in the ductile matrix, and as a result these composites exhibit extremely low ductility.

The key idea is to apply sufficient shear stress on particle clusters embedded in the liquid metal to overcome the average cohesive force or the tensile strength of the cluster. Molecular dynamics studies [12–14] suggest that the intensive shearing can displace the position of atoms that are held together with high-strength bonds. Under a high shear and high intensity of turbulence, liquid can penetrate into the clusters and displace the individual particles within the cluster. A new Rheo process is introduced for the production of AlSiC PMMCs utilizing the melt conditioning by advanced shear technology (MCAST) process developed at the Brunel Centre for Advanced Solidification Technology (BCAST), Brunel University, London, in which the liquid undergoes a high shear stress and high intensity of turbulence inside a specially designed twin-screw machine. Adetailed description of process was reported by S. Tzamtzis [15].

Typical microstructures can be seen in Fig. 8.2a. The microstructures obtained after dispersive mixing show a more uniform distribution compared with the distributive mixing process. The melt conditioned (MC) cast samples have high levels of porosity similar to the compo-cast ones. This is due to the fact that the distributive mixing process, prior to the Rheo processing of the composites, causes the suction of air bubbles, as mentioned previously. The casting method also induces porosity in the samples. However, for the MC- high-pressure die casting (MC-HPDC) samples the resulting microstructures were highly improved. The uniform dispersion of SiC particles in the matrix is clearly seen in Fig. 8.2b.

The samples produced with the Rheo process have a much more homogeneous microstructure, the particles being distributed more uniformly throughout the whole volume of the sample. This was consequently reflected in the mechanical properties acquired by the tensile tests carried out. Figure 8.3 shows a comparison of the mechanical properties of A356/SiC composite samples produced by the two different processes. The MC-HPDC composites show an increase in the tensile elongation together with an increase in the ultimate tensile strength (UTS) of the material, resulting from the effective dispersion of the particles. The magnitude of this increase is about 15%.

8.3.3 Modified Squeeze Casting Technique

Another significant challenge for the processing of aluminum-based thermal management materials is to fabricate AMCs with a high volume fraction of SiC, which has an average diameter of less than 10 μm. Shoujiang Qu et al., of the Harbin Institute of Technology in China, report a novel manufacturing process that overcomes this limitation [16].

Particle reinforced aluminum composites (PRACs) are usually fabricated by a casting process and powder metallurgy (PM) technique. With the use of the PM technique, PRACs with various sizes and volume fractions of the particles can be fabricated easily. However, the complicated processes and relatively high cost limit wide application of PM products. Casting processes such as squeeze casting and stir casting are cost effective. In general, particles with an average diameter of less than 10 μm or a volume fraction above 20% are extremely difficult to introduce into a melt. SiC particles with an average diameter of 3.5 μm are a commercially viable, low-cost material for reinforcing aluminum.

A cost-effective method was introduced to fabricate pure AMCs reinforced with 20% volume fraction of 3.5 m SiC particles by squeeze casting followed by hot extrusion. To lower the volume fraction of the composites, a mixed preform containing pure aluminum powder and SiC particles was used. The suitable processing parameters for the infiltration of pure aluminum melt into a mixed preform are as follows: melt temperature 800°C, preform temperature 500°C, infiltration pressure 5 MPa, and

Fig. 8.2 Typical microstructures of A356 – 5 vol% SiCp (4 μm particle size). (**a**) MC-cast. (**b**) MC-HPDC

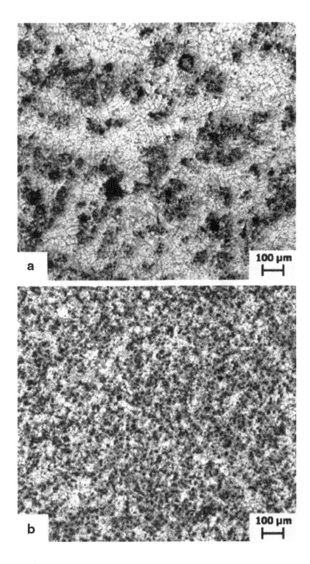

solidification pressure 50 MPa. The microstructure and properties of composites in both as-cast and hot-extruded states were investigated. The results indicate that hot extrusion can obviously improve the mechanical properties of the composites.

One of the most significant challenges for the processing of aluminum-based thermal management materials is the formation of aluminum oxide at high temperature with the presentation of air at high temperature. Thus the central theme in the manufacturing process is to isolate air from the process by utilizing a vacuum chamber. It has been discovered that the formation of oxide is inevitable (Fig. 8.4a), and extrusion could break the oxide into pieces. However, TEM images reveal that total elimination is difficult to achieve.

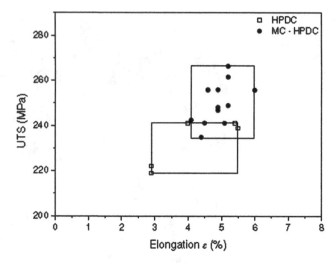

Fig. 8.3 Comparison of mechanical properties of A356/SiCp 4 μm composites obtained from different processes

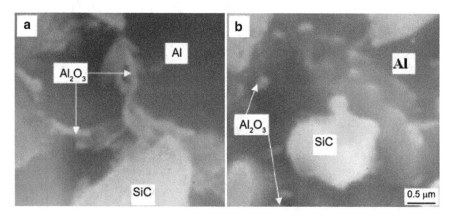

Fig. 8.4 SEM photos showing the morphology of the Al2O3 phases in the SiCp/Al composites. (**a**) As cast. (**b**) As extruded

The SEM photos in Fig. 8.4 show the morphology of the Al_2O_3 phases in the SiCp/Al composites: Fig. 8.4a as cast, Fig. 8.4b as extruded. However, the morphology of the Al_2O_3 phases both in as-cast and as-extruded composites was also observed by the TEM (Fig. 8.5). The composition analysis using an EELS indicates the existence of the Al_2O_3 phases.

Aluminum oxide formed at the surface of aluminum is undesirable because it reduces the tensile strength and ductility. Once the oxide is formed, it can be broken

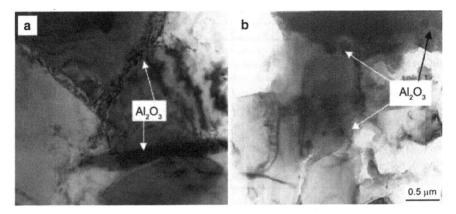

Fig. 8.5 TEM images showing the existence of the Al2O3 phases in SiCp/Al composites. (**a**) As cast. (**b**) As extruded

Table 8.4 Tensile properties of SiCp/Al composites

Material	Tensile strength/MPa	Modulus/GPa	Elongation/%
As-cast composite	140	75	0.5
As-extruded composite	240	90	3.0

into small pieces by extrusion. The mechanical properties such as tensile strength and ductility of such materials can be improved accordingly (Table 8.4).

8.3.4 Aluminum Silicon Carbide (AlSiC) for Advanced Microelectronic Packages

A leader in developing AlSiC for thermal management applications is Thermal Transfer Composites based in Hockessin, DE. The properties of typical AlSiC materials shown in Table 8.5 and product photos shown in Fig. 8.6 are used with their permission.

8.3.5 Effects of Temperature on Pressure Infiltration of Liquid Aluminum into Packed SiC Particles

It is well known that a minimum external pressure is required to infiltrate a liquid metal into porous media in nonwetting conditions. The temperature at which infiltration of ceramic preforms by liquid metals is carried out is one of the key

Table 8.5 Properties of typical AlSiC materials by Thermal Transfer Composites

Property data	AlSiC-1	AlSiC-2	AlSiC-3	AlSiC-4	AlSiC-5	AlSiC-6
Al wt%	63	55	45	42	40	22
SiC wt%	37	45	55	58	60	78
TC (W/mK) at 25°C	180	165	180	185	192	255
CTE (ppm/K)	10.9	7.2	6.4	7	6.9	4.8
Density (g/cm³)	2.9	2.94	2.98	3	3.01	3.1
Young's modulus (GPa)	188	235	255	240	262	342
Poisson's ratio	–	0.23	0.24	0.23	0.25	0.23
Flexural strength, 4 pt bend (MPa)	–	330	300	330	517	225
Specific heat at 25°C (J/gK)	0.81	0.74	0.74	0.74	0.74	0.71
Electrical resistivity (μΩ.cm)	20.7	30–50	30–50	30–50	30–50	20–25

experimental variables in the fabrication of composites by means of pressure infiltration. The threshold pressure P_0 for infiltration of pure aluminum into a variety of ceramic preforms decreases with temperature.

Tian [17] reports on the effects of temperature on pressure infiltration of packed SiC particles by liquid Al and the eutectic Al–12 wt% Si alloy over the temperature range 923–1,273 K (Figs. 8.7 and 8.8). Below 1,173 K, the threshold pressure P_0 for infiltration of pure Al slightly (and linearly) decreases with temperature. At that temperature a sharp drop occurs, followed by a faster decrease of P_0. The contact angle h derived from the threshold pressure data shows a similar behavior.

8.3.6   High-Temperature (>200°C) Packaging of SiC Power Converters Using Aluminum Silicon Carbide (AlSiC)

Silicon devices are limited to a 150°C junction temperature before derating, where SiC devices can operate in excess of 400°C. Hopkins demonstrates that it is feasible for a 60 kW air-cooled SiC, 3-phase converter to operate at 0.1 MHz using an AlSiC heat sink [18]. The converter operates from a 650 V dc bus and in an ambient temperature of 150°C and lower. The converter can deliver 150 A peak/60 kW from a 650 V dc with a density of 1.1 kW/in.³. The converter operates from a 650 V dc bus in an ambient temperature of 150°C. The primary SiC devices and packaging can operate continuously at 350°C with high reliability.

This converter design is unique in that the mechanical reliability preceded the electrical design. A paramount requirement is to minimize the dissimilarity in material interfaces. A "nearly all" Al (aluminum) approach is taken to optimize design and manufacturability. The AlSiC shown in Fig. 8.9 functions as a heat sink. The top ten pieces are SiC devices. The next structures are aluminum conductors, the third structure is an Al/Si_2O_3 ceramic insulator, and the fourth structure is a heat sink made of AlSiC.

Fig. 8.6 Pictures of AlSi parts. (**a**) AlSiC IGBT base plates. (**b**) AlSiC liquid-cooled components. (**c**) AlSiC air-cooled components. (**d**) AlSiC packages by Thermal Transfer Composites (www. thermaltc.com)

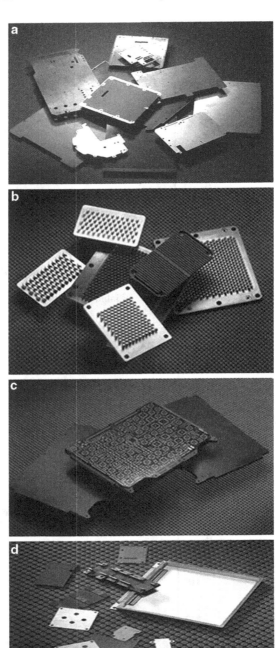

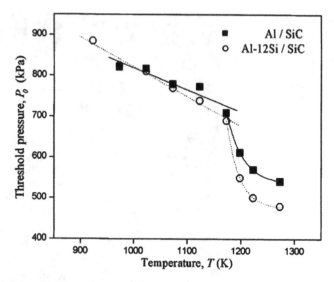

Fig. 8.7 Threshold pressure P_0 for infiltration of liquid Al and Al–12 wt% Si alloy into compacts of SiC particles versus temperature T

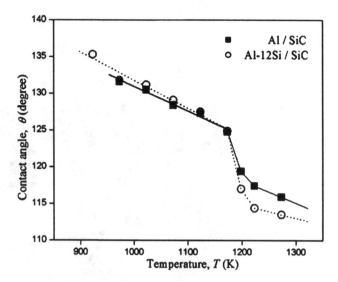

Fig. 8.8 Contact angle h derived from experimental results for threshold pressure for infiltration of Al and Al–12 wt% Si into SiC compacts versus temperature T

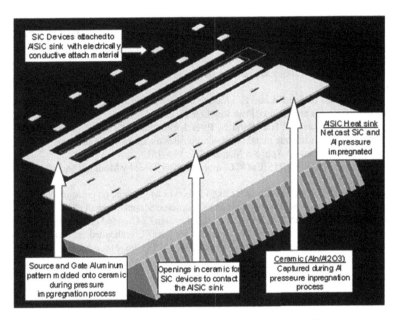

Fig. 8.9 Heat sink layout for inverter. The top ten pieces are SiC devices. The next structures are aluminum conductor, the third structure is an Al/Si$_2$O$_3$ ceramic insulator, and the forth structure is a heat sink made of AlSiC

References

1. Zweben C (2006) Thermal materials solve power electronics challenges. Power Electr Technol 32:40
2. Occhionero MA, Hay RA, Adams RW, Fennessy KP (2011) Aluminum silicon carbide (AlSiC) for cost-effective thermal management and functional microelectronic packaging design solutions. http://cpstechnologies.wsiefusion.net/pdf/cpseuro992.pdf
3. Surappa MK (2003) Aluminium matrix composites: challenges and opportunities. Sadhana 28:319–334
4. Lloyd DJ (1999) Particle reinforced aluminium and magnesium matrix composites. Int Mater Rev 39:1–23
5. Surappa MK, Rohatgi PK (1981) Preparation and properties of aluminium alloy ceramic particle composites. J Mater Sci 16:983–993
6. Molina JM, Saravanan RA, Arpon R, Garcia-Cordovilla C, Louis E, Narciso J (2001) Pressure infiltration of liquid aluminium into packed ceramic particulate with a bimodal size distribution. Trans JWRI 30(Special Issue):449–454, Join Weld Res Inst Osaka University
7. Segurado J, Gonzáles C, Llorca J (2003) A numerical investigation of the effect of particle clustering on the mechanical properties of composites. Acta Mater 51:2355–2369
8. Deng X, Chawla N (2006) Modeling the effect of particle clustering on the mechanical behavior of SiC particle reinforced Al matrix composites. J Mater Sci 41:5731–5734
9. Lloyd DJ (1991) Aspects of fracture in particulate reinforced metal matrix composites. Acta Metall Mater 39(1):59–71
10. Nair SV, Tien JK, Bates RC (1985) SiC reinforced aluminium metal matrix composites. Int Met Rev 30(6):275–290

11. Youssef YM, Dashwood RJ, Lee PD (2005) Effect of clustering on particle pushing and solidification behavior in TiB2 reinforced aluminium PMMCs. Compos Part A 36:747–763
12. Falk ML, Langer JS (1998) Dynamics of viscoplastic deformation in amorphous solids. Phys Rev E 57:7192–7205
13. Schuh CA, Lund AC (2003) Atomistic basis for the plastic yield criterion of metallic glass. Nat Mater 2:449–452
14. Mayr SG (2006) Activation energy of shear transformation zones - a key for understanding rheology of glasses and liquids. Phys Rev Lett 97:195501
15. Tzamtzis S, Barekar NS, Hari Babu N, Patel J, Dhindaw BK, Fan Z (2009) Processing of advanced AlSiC particulate metal matrix composites under intensive shearing – a novel Rheo process. Compos Part A: Appl Sci Manuf 40(2):144–151
16. S Qu, Geng L, Han J (2007) SiCp/Al Composites Fabricated by Modified Squeeze. J Mater Sci Technol 23(5)
17. Tiana J, Piñerob E, Narciso J, Louis E (2005) Effects of temperature on pressure infiltration of liquid Al and Al–12 wt% Si alloy into packed SiC particles. Scripta Materialia 53(12):1483–1488
18. Hopkins DC, Kellerman DW, Wunderlich RA, Basaran C, Gomez J (2006) High-temperature, high-density packaging of a 60 kW converter for >200°C embedded operation. In: Applied power electronics conference and exposition, 2006. APEC '06. 21st Annual IEEE, Dallas, TX, 19–23 Mar 2006

Chapter 9
Understanding Lasers, Laser Diodes, Laser Diode Packaging and Their Relationship to Tungsten Copper

Abstract The backbone of modern telecommunications is optical fibers. For each end of every optical fiber, there is one laser or laser diode transmitting the light and one photo detector to decode the signals. As Internet traffic expands and the transmission distance increases, more and more designs are using higher and higher power lasers or laser diodes. Once again this development calls for better thermal management designs and better thermal management materials. This chapter serves as a layman's introduction to lasers, laser diodes, and laser diode packaging. Within the thermal management scope, the use of copper tungsten is examined in detail. As technology evolves, the introduction of newer and more advanced material will take place of copper tungsten materials.

9.1 What Is a Laser?

Light amplification by stimulated emission of radiation, or laser for short, is a device that creates and amplifies electromagnetic radiation of specific frequency through a process of stimulated emission. In laser, all the light rays have the same wavelength and are coherent; they can travel long distances without diffusing.

To understand how lasers work, we must understand how an atom gives out light. An atom is among the smallest particles in the world, and it contains electrons. When extra photons are introduced into an atom, the atom's electrons are forced to move to a higher energy level, and now the atom is in an excited state. However, the excited atom is unstable, and the electrons try to return to their ground state, thereby releasing as photons of light radiation the excess energy they originally gained. This process is called spontaneous emission (Fig. 9.1).

The laser contains a chamber in which atoms of a medium are excited, bringing their electrons into higher orbits with higher energy states. When one of these electrons jumps down to a lower energy state, it gives off its extra energy as a photon with a specific frequency. When more photons are introduced into the system, the

G. Jiang et al., *Advanced Thermal Management Materials*,
DOI 10.1007/978-1-4614-1963-1_9, © Springer Science+Business Media New York 2013

Fig. 9.1 Emission of photon
as light energy

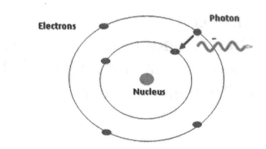

Fig. 9.2 Stimulated emission
of photons

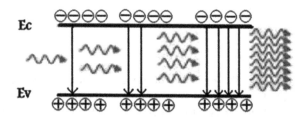

photons will eventually encounter another atom with an excited electron, which will stimulate that electron to jump back to its original state, emitting two or more photons with the same frequency as the first and in phase with it. This effect cascades through the chamber, constantly stimulating other atoms to emit yet more coherent photons, and this process is called stimulated emissions. In other words, the light has been amplified, as shown below in Fig. 9.2.

Furthermore, mirrors at both ends of the chamber cause the light to bounce back and forth across the medium. One of the mirrors is partially transparent, allowing the laser beam to exit from that end of the chamber. By maintaining a sufficient number of atoms in the medium by an external energy source in the higher energy state, the emissions are continuously stimulated, and this process is called population inversion. Ultimately, it creates a stream of coherent photons, which is a very concentrated beam of powerful laser light. Lasers have many industrial, military, and scientific uses, including welding, target detection, microscopic photography, fiber optics, surgery, and others.

9.2   Types of Laser

There are many different types of laser. What follows is a list of the five major types.

1. Gas lasers – e.g., HeNe gas laser and CO_2 lasers that emit hundreds of watts of power. They are usually used for cutting and welding in industry.
2. Chemical lasers – powered by chemical reaction, which permits large amounts of energy, mainly for military use and of very high wavelength; e.g., 2,700 nm hydrogen fluoride laser.

3. Solid-state lasers – optically pumped using a solid medium that is doped, such as ion-doped crystalline or glass. An example would be a laser pointer.
4. Fiber lasers – light is guided by internal reflection in an optical fiber. These lasers are widely known today for their high output power and high optical quality as well as long life span. This is due to the properties of fibers that have a high surface-area-to-volume ratio, which allows for efficient cooling when supporting kilowatts of continuous output power. A fiber's wave-guiding properties help maintain signal strength and minimize distortion. Today, fiber lasers are widely used for telecommunications that spread across regions several kilometers long.
5. Semiconductor lasers – electrically pumped

   (a) Light-emitting diodes (LEDs) - in a diode formed from a direct band-gap semiconductor, such as gallium arsenide, carriers that cross a junction emit photons when they recombine with the majority carrier on the other side. Depending on the material, wavelengths (or colors) from the infrared to the near ultraviolet may be produced. All LEDs produce incoherent, narrow-spectrum light. LEDs can also be used as low-efficiency photodiodes in signal applications. An LED may be paired with a photodiode or phototransistor in the same package to form an opto-isolator.
   (b) Laser diodes - when an LED-like structure is contained in a resonant cavity formed by polishing the parallel end faces, a laser can be formed. Laser diodes are commonly used in optical storage devices and for high-speed optical communication.

9.3   A Closer Look at Laser Diodes

A laser diode is a laser where the medium is a semiconductor formed by a p-n junction (Fig. 9.3) and powered by electric current. For different types of laser diode structures, see Appendix 3. Basically, a laser diode is a combination of a semiconductor chip that emits coherent light and a monitor photodiode chip for feedback control of power output in a hermetically packaged and sealed case.

The semiconductor materials that are used to create p-n junction diodes that emit light today are g, indium phosphide, gallium antimonide, and gallium nitride. The reason that these are used is because of the three-five compound properties in the periodic table of elements. The materials must be heavily doped to create P – N regions, which rules out others, leaving the three-five groups with the ideal options.

Table 9.1 gives several typical substrate materials used in various laser applications.

Their wavelengths can be adjusted by changing the ratio of composition. For instance, the wavelength of a laser beam produced by an InP substrate can be increased by increasing the indium content or lowering the phosphate content percentage. A longer wavelength usually indicates a longer travel distance.

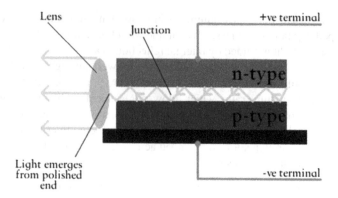

Fig. 9.3 A laser diode converts electrical energy into energy in the form of light (credit to Explainthatstuff.com)

Table 9.1 Laser substrates, their wavelength ranges, and applications

Laser diode substrate material	Typical wavelength (nm)	Applications
GaN/InGaN	350–500 (UV - blue)	Blu-ray disc, biomedical fluorescence
GaAlAs/GaAs	530–980 (red, near-infrared)	DVD, CD, biomedical, printing, industrial aligning, sensing, cosmetics, laser projection, bioanalysis (spectroscopy)
InP/InGaAsP	900–1,650 (infrared)	Range-finder, fiber optics, telecommunication, data transmissions, network, optical pumps

Laser diodes are numerically the most common type of laser, with 2004 sales of approximately 733 million units, as compared to 131,000 of other types of lasers. Laser diodes find wide use in telecommunications as easily modulated and easily coupled light sources for fiber-optics communication.

9.3.1   Laser Packaging

To protect laser diode materials or any laser devices from mechanical and thermal stress, because the laser material, for example, gallium arsenide is very fragile, laser packaging is required for virtually all laser diodes or other laser devices. If we imagine the laser diode as a pizza, the packaging mount serves the purpose of pizza box that holds the pizza inside.

Additionally, the hermetic package sealing method prevents dust or other contaminations from entering the laser; smoke, dust, or oil can cause immediate or permanent damage to a laser. Most importantly, as technologies advance, the

Fig. 9.4 Integration of submount in laser packaging

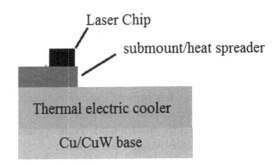

emergence of high-power diode lasers requires a sophisticated packaging design to help release the heat dissipated during operation through submounts and mounted heat sinks.

Usually, the laser die is mounted on a submount of material with a similar coefficient of thermal expansion (CTE) that matches the CTE of the die. The submount serves as a heat spreader, which gets further mounted onto a thermal electric cooler (TEC) for rapid heat transferal. The entire assembly is then mounted onto a pure copper or copper/tungsten heat sink for advanced heat dissipation and cooling (Fig. 9.4).

High-power laser diode or laser devices convert electrical energy into light energy at about 10–50 % efficiency. The remaining energy is generated as waste heat and must be dissipated within a short amount of time or else it will cause thermal stress on the laser diode bar and eventually cause irreversible damage to the laser.

Inefficient cooling packaging design will result in poor product quality as the temperature of the device core has a direct influence on output wavelength and band gap. It has been proven in practical situations that for every 3 °C of change, the wavelength of the diode laser can change by nearly 1 nm. Also, the output power of the laser will decrease as temperature increases.

To help dissipate the heat, the conventional method of laser diode packaging is to solder a laser bar onto a heat sink that is made of copper due to its high thermal conductivity (TC), that is, faster heat transfer. Traditionally, soft indium solder is used to bond a heat sink to a laser material like gallium arsenide (GaAs) because the CTE of copper does not match that of the laser material well enough. Indium has a higher ductility than copper and provides greater reliability for continuous wave (CW) and quasi-continuous wave (QCW) operations.

Definition of Continuous Wave (CW) and Quasi-CW:

1. CW – continuous wave operation of a laser means that the laser is continuously pumped and continuously emits light. The emission can occur in a single resonator mode (→ *single-frequency operation*) or in multiple modes.
2. QCW (pulsed) - In continuous-wave operation, some lasers exhibit too strong heating of the gain medium. The heating can then be reduced by QCW operation,

where the pump power is only switched on for limited time intervals. QCW operation of a laser means that its pump source is switched on only for certain time intervals that are short enough to reduce thermal effects significantly but still long enough that the laser process is close to its steady state, i.e., the laser is optically in the state of CW operation. The duty cycle (percentage of "on" time) may be, e.g., a few percent, thus strongly reducing the heating and all the related thermal effects, such as thermal lensing and damage through overheating. Therefore, QCW operation allows an operation with higher output peak powers at the expense of lower average power.

A pulsed operation with significantly shorter pumping times, where an optical steady state is not reached, is called *gain switching*.

QCW operation is most often used with diode bars and diode stacks. Such devices are sometimes even designed specifically for QCW operation: their cooling arrangement is designed for a smaller heat load, and the emitters can be more closely packed in order to obtain a higher brightness and beam quality. Compared with ordinary CW operation, additional lifetime issues can result from QCW operation, related, e.g., to higher optical peak intensities or to frequent temperature changes.

Some doped-insulator solid-state lasers are also operated in QCW operation. Such lasers are sometimes called *heat capacity lasers*.

QCW operates at typically higher frequency kHz (ns-ms pulse) compared to pulsed Hz. Mostly, QCW is used to extend battery life or reduce heat. QCW and pulsed Hz are usually implemented by adding a laser diode driver to control the input for different applications.

However, the repeat on-off cycle/hard pulse in laser operations can cause mechanical stress, which leads to material cracking/indium migration and, further, to failure. It happens in direct diodes and solid-state pump diodes. Usually, low-power lasers can last longer before quitting. High-power lasers often encounter such issues and fail much quicker, as they have much larger contacting surfaces between die and heat sink; thermal expansion of heat sinks is always a major issue.

With increasingly advanced technology, many laser R&D companies have developed more sophisticated high-power and high-performance diode laser arrays suitable for both CW and QCW operations. CW and QCW powers in excess of 900 W have been demonstrated at various laser companies, e.g., Princeton Optronics. Yet, the mounting of these laser bars still has not changed much in performance in terms of, e.g., heat conduction, resistance, and expansion. While the higher thermal conductive copper heat sink submount offers a cost-effective solution, the mismatching thermal expansion rate with laser materials remains an issue, and the problem is being magnified as devices start to produce more heat. The reason for this is that the material property of copper, which tends to expand as temperature increases, causing mechanical stress on the laser diode. Apparently, a pure copper heat sink can no longer keep these products within thermal expansion tolerance.

9.3.2 Tungsten Copper as Submount and Heat Sink

Tungsten copper, alternatively, provides a much lower thermal expansion rate compared to pure copper while maintaining a necessary thermal conduction rate. Tungsten copper is a copper and tungsten alloy that usually consists of 10–50 % of its weight in copper and the remaining portion in tungsten. The higher the tungsten content, the lower its thermal expansion rate is.

The copper/tungsten heat sink submount with gold/tin solder is a very cost-effective method that provides both good TC and thermal expansion that matches those of the silicon ceramic and gallium arsenide materials used to make circuits for semiconductors and for both high-power and low-power laser diodes and bars.

Shown below are comparison graphs with descriptions by Dr. Yuen from coherent.com. One can clearly see that the tungsten copper significantly helped increase the laser die chip lifetime during operations (Fig. 9.5).

The current trend of increasing die size and power dissipation requirements has made CuW the ideal material of choice for laser diode packaging. In addition, the conventional CuW heat sink submount provides a TC of 180–230 W/mK with a CTE of 6.5–9.0 ppm/deg. C that matches the die of laser diodes. And with newly improved solutions such as the finite boundary value method and functionally graded materials, the TC of copper tungsten can be pushed up to approximately 320 W/mK. All of these thermal management solutions can be pursued and achieved using tungsten copper, which is one of the most common, readily available materials.

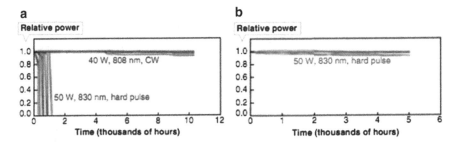

Fig. 9.5 The lifetime is seriously degraded in QCW/pulse laser operations (**a**), while the lifetime is significantly better using an improved submount made from CuW (**b**)

9.3.3   Tungsten Copper as Submount and Heat Sink in High-Power Diode Arrays

The technologies for high-power diode lasers have been rapidly developing in recent years; however, packaging technology remains a bottleneck for the advancement of high-power semiconductor lasers.

In addition to the usage of a CuW submount in single-emitter laser packaging applications, tungsten copper is also a thermal management solution for high-power diode lasers, which are created by combining several single emitters in an array. These high-power diode arrays are applicable in pumping solid-state laser systems for industrial, commercial, military, and medical applications as well as material processing applications such as welding, cutting, and surface treatment.

Shown below in Fig. 9.6 is a picture comparison between a single emitter and an array of emitters.

Clearly, the array is a combination of multiple single laser diodes; therefore, more power and heat are produced. Also, the multiple beams emitted form a near field and its linearity (or "smile") are an important parameter in determining the overall coupling efficiency between the diode lasers and the fiber or optic lens. The near-field linearity depends largely on the degree of CTE mismatch between the die and bonding materials, which will be discussed later.

In simple words, the performance and longevity of the laser array depend largely on the thermal management of the laser package. To achieve high efficiency and high power, the heat sink submount must have the capability of transferring heat at a very fast speed while maintaining a relatively close thermal expansion rate with the die material. Tungsten copper is often used in situations like this due to its high TC and good CTE match with that of die material (GaAs) and to its electrical active property of acting as a P-side. The CuW submount is often bonded to a pure copper or thermal electric cooler for advanced heat dissipation.

9.3.4   CuW in Comparison with AlN and BeO in Submounts

Typically, CuW, AlN, and BeO are used as submounts in packaging laser diodes. The CuW material is a metal composite, while the AlN and BeO composites are

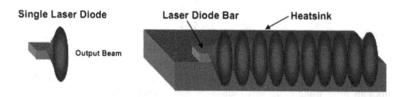

Fig. 9.6 Image of diode laser bars consisting of multiple single emitters on a single substrate (from Coherent.com)

Table 9.2 Advantages and disadvantages of three composite materials

	CuW	AlN	BeO
CTE (ppm/deg. C)	6.5–9.0	4.5	7
TC (W/mK)	180–230	170–200	280
Density (g/cm³)	14.9–17	3.3	2.9
Material information	Metal composite, electrically conductive, can act as P-side in P-N junction, strong and durable, CTE matches well that of GaAs	Ceramic, electric insulation, good high-frequency response[a], CTE matches well that of InP, easy for gold plating (tracing) for programming diode	Ceramic, electric insulation, high TC, CTE matches well that of GaAs, toxic

[a]AlN is suitable for high-frequency wave applications due to its ceramic material property; it helps minimize signal distortion and interference

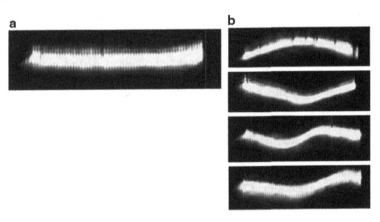

Fig. 9.7 (a) Enlarged "smile" of a good diode-laser array. (b) Examples of various "smiles" due to CTE mismatch between diode-laser arrays and bonding submount/heat sinks

ceramic; they serve different purposes. The CuW composite is typically used for heat-spreading purposes, while AlN and BeO are dielectric materials that are used for electrical insulation purposes. Table 9.2 compares the pluses and minuses of each material selection.

The CTE plays an important role in determining the overall performance and degradation time of diode laser products. Materials expand in size due to increasing temperature. A slight degree of CTE mismatch could cause a "smile," which will result in poor laser performance. A "smile" is a nonlinearity of the near field of emitters. Shown below on the left is an image of a typical good diode-laser array in comparison with images of diode-laser arrays that have various "smiles" from Dr. Liu's experiment (Fig. 9.7).

A high degree of CTE mismatch between bonding materials can cause a laser die to crack during sintering or brazing. To prevent such a catastrophe, the CTE of bonding layers should always have a close match to the that of the laser die. Sophisticated thermal management helps the device maintain a strong signal and better product quality.

9.4   Types of Laser Diode Packaging that Require CuW Material

1. C-mount packages
   These are used for lasers and laser-based systems, laser measurement and control, and precision optics. Typical wavelength is from 680 to 980 nm, with output power rating up to 7 W (Fig. 9.8).
   The laser diode is directly soldered onto the copper-tungsten heat sink, which acts as a P-side, and the other side of the diode cavity is a wire bonded to a metal contact, which is the cathode. The hole in the middle is used for mounting purposes.
2. TO3 package
   There are six to eight pins. The base material is either CRS or CuW (Fig. 9.9).
3. VCSEL submodule package
   The vertical-cavity surface-emitting diode lasers are have become popular over the past three decades due to their low cost and high reliability.
   They are a great choice for short-range data communications and networks.
   Their output power and wavelength vary depending on application requirements.
   Shown below on the right is the architecture of a VCSEL; the P-contact is a submount that is usually made of heat sink materials that have high TC whose CTE closely matches the CTE of a GaAs die substrate such as CuW or CuMo (Fig. 9.10).

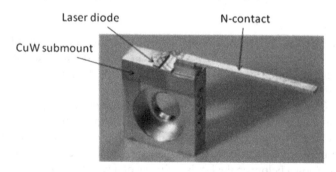

Fig. 9.8   C-mount laser diode

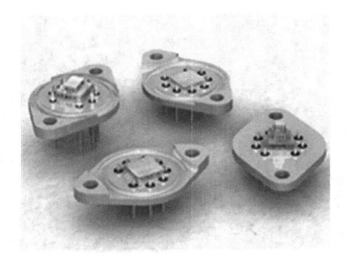

Fig. 9.9 Image of TO3 packages

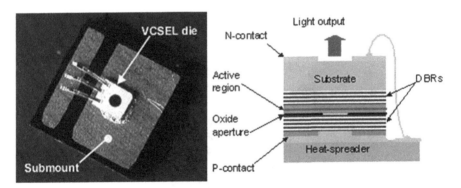

Fig. 9.10 On the left is an image of a high-power VCSEL submodule package from Princeton Optronics. Submount is 2×2 mm and output power is >2 W

4. BTF (Butterfly) package

   The butterfly package is the standard format for optical telecom transmissions and laser diode pumps. Below is a typical 14-pin butterfly package, in which the laser die sits on an AlN submount. The AlN submount is mounted on a TEC, which is attached to a baseplate made of CuW, Kovar, or CuMo (Fig. 9.11).

5. Mini-DIL package

   The dual in-line (DIL) package is promising for telecomm applications and usually has six to eight pins. The base can be various: CuW, CRS, or alumina (Fig. 9.12).

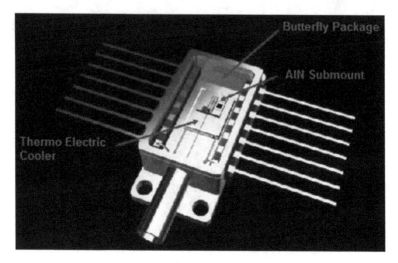

Fig. 9.11 Sample butterfly package

Fig. 9.12 Mini-DIL package

6. TOSA package

   TOSA/ROSA packages or transmitter/receiver optical subassemblies are mainly for use with transceivers and transponders for data transmission purposes. Shown below is an example from Hitachi-Hitech. Their bases are typically made of copper tungsten to accommodate heat dissipation (Fig. 9.13).

7. HHL package

   High heat load (HHL) packages are the largest standard laser diode packages available. They are designed for high-power diode laser applications and usually have nine pins. The base material can be various: Kovar, CuW, or CuMo (Fig. 9.14).

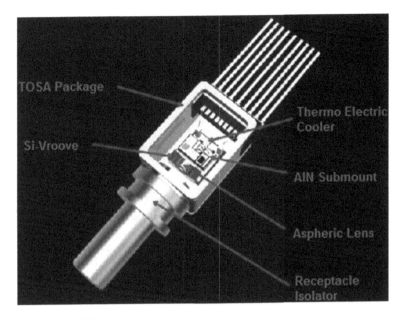

Fig. 9.13 Sample TOSA package

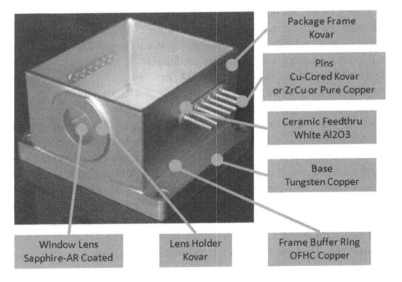

Fig. 9.14 Architecture of HHL package

8. Golden/Silver Bullet Laser Array Submodules (ASM package)
   Application: Typically used for solid-state laser pumping
   Typical wavelength: 803–808, 880, 885, or 940 nm (golden bullet package)
      800–1,550 nm (silver bullet package)

Fig. 9.15 *Top*: golden bullet packages; *bottom*: silver bullet packages. There are one-bar, two-bar, and three-bar designs

Output power: 20–40 W (CW), 50–300 W (pulsed)

The base is usually of ceramic coolers, whereas the end blocks on both sides are made of CuW for both thermal and electrical conduction (Fig. 9.15).

9. CCP package (CS-mount)

Application: Conduction-cooled packaging (CCP) method for diode laser bar, used in laser systems or direct-diode application.

Typical wavelength: 806, 880, 885, or 940 nm

Output power: 20–40 W (CW), 100–1,600 W (pulsed)

Typical material layer schematics:

Shown above is an example of laser die bonding in a CS-type mount; a CTE-matched submount is added between the die and the heat sink to serve as a buffer layer. The diode bar is bonded to a CuW submount using gold/tin solder, which is mounted on a copper heat sink, acting as the P-side. The N-side is wire-bonded to the P-side. The typical output power is 60–300 W in either CW or QCW mode, making it well suited for solid-state pumping, direct-diode material processing, medical, reprographics, and illumination applications (Fig. 9.16).

10. CCP stacks (G-mount)

Applications: The high output power is suitable for military, range finding, sensing, and medical applications.

Typical wavelength: 808, 880, 885, or 940 nm

Output power: 20 W (CW), 100–5,200 W (pulsed)

Below is the architecture of Coherent's Vertical diode laser array. Each diode bar is soldered to the CuW submount with AuSn solder on the P-side. The N-side is connected with indium solder (Fig. 9.17).

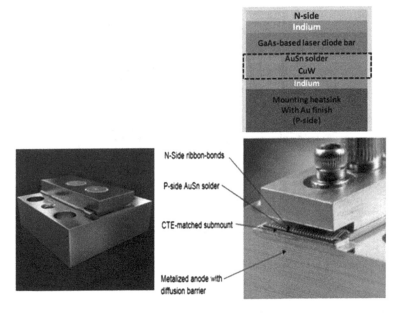

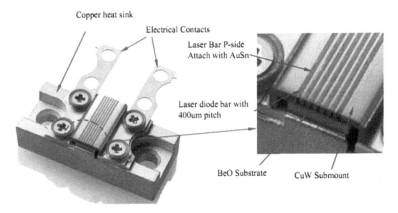

Fig. 9.16 Schematics of CS-mount

Fig. 9.17 Architecture of Coherent's 7-bar vertical array. Maximum peak power: 1,400 W (200 W/bar)

The submount is made of CuW composite due to its perfect thermal expansion match with the laser diode material (GaAs). The BeO layer allows for quick heat transferal onto the copper heat sink, while helping to maintain electrical insulation between the diode and the heat sink.

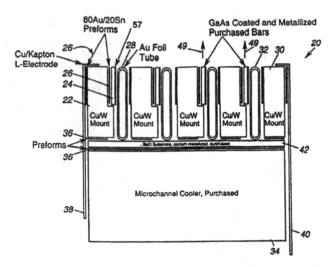

Fig. 9.18 Microchannel cooled diode-laser array

11. Microchannel cooled package (MCCP)

Application: The high output power achieved by these packages can be used in laser pumping, military (range finding, light detecting), or medical applications.

Typical wavelength: 806–808, 880, or 980 nm

Output Power: 40–250 W

Due to the excellent thermal and electrical conduction of the CuW submount and the close CTE match with that of the die material GaAs, the MCC package allows stacking of several high-power laser diodes in an array using gold/tin solder, and the subassembly is then cooled using water cooling channels, as shown below (Figs. 9.18 and 9.19).

9.5   Conclusion

CuW has been used recently in many places in laser packaging, especially in die chip submounts and heat sinks. Other techniques of joining copper tungsten to other metals such as Kovar to form miniature heat sinks, subcarriers, or subassemblies suit applications that require lightweight design. As higher power laser diode devices emerge, the requirements for diode laser mounting will increase. Due to the current technology trend in developing higher power diode lasers, the mounting substrate or heat sink has a significant impact on the performance of the diode laser system. Materials that were previously used cannot satisfy the thermal management requirement of diode lasers nowadays. Fortunately, copper tungsten provides high reliability for enclosing electronic material from the outside environment and provided an improved TC for mounting and integrating high-power laser diode devices.

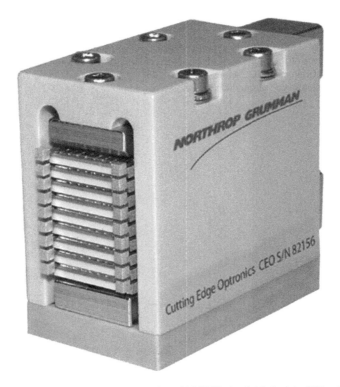

Fig. 9.19 Example of a 6-bar MCC package from NG/CEO. Available both in CW or QCW bars with maximum power rating of 1,800 W

References

1. Yuen A (2007) Telecom packaging improves reliability of high-power lasers. Laser Focus World, Semiconductor Resources. (www.coherent.com)
2. Bhatia R (2003) Materials issues and engineering design considerations for device packaging of high power edge emitting semiconductor laser arrays and monolithic stacked laser diode bars. Department of Chemical & Materials Engineering, San Jose State University, California
3. Liu X, Zhao W (2009) Technology trend and challenges in high power semiconductor laser packaging. State Key Laboratory of Transient Optics and Photonics, Xi'an Institute of Optics and Precision Mechanics, Chinese Academy of Sciences, China
4. Wang J, Yuan Z et al (2009) Study of the mechanism of "smile" in high power diode laser arrays and strategies in improving near-field linearity. State Key Laboratory of Transient Optics and Photonics, Xi'an Institute of Optics and Precision Mechanics, Chinese Academy of Sciences, China
5. Sepulveda JL, Valenzuela L et al (2000) Copper/tungsten mounts keep diode lasers cool. Optoelectron Packaging, Telecom Products Division, Opto Power Corp
6. Rohlin A (2011) What is tungsten copper use for? www.ehow.com
7. CW and QCW definitions (2011) http://www.rp-photonics.com
8. Boucke K, Jandeleit J et al (2000) Packaging and characterization equipment for high-power diode laser bars and VCSELs. University of Technology, Germany
9. CVI Melles Griot (2011) All things photonics, vol 2. Issue 1, pp 10.1–10.32

10. L.E.W. (2002) Techniques Press Releases Archive http://www.lewtec.co.uk/ecoc_press_release.htm
11. Kyocera (2011) http://americas.kyocera.com/kai/semiparts/products/index.cfm
12. Hitachi Hitech (2011) http://www.hitachi-hitec.com/global/oc/index.html
13. Northrop Grumman/CEO Laser Diodes (2011) http://www.as.northropgrumman.com/businessventures/ceolaser/products/laserdiodes/index.html

Chapter 10
Future Trend of Advanced Thermal Management Materials

Abstract The R&D work in advanced thermal management materials is active and ongoing. It is difficult to cover all related research results. In this chapter, we select a few common materials used in microelectronics packaging to review future trends. While most material is typically of a homogeneous structure, Cu/Mo70Cu/Cu is not. As discussed in an earlier chapter, this material (typically at 1:4:1 ratio) has a coefficient of thermal expansion in the X-direction of 7.2 ppm/°C and 9.0 ppm/°C. It was used successfully for long and this die attaches such as laterally diffused metal-oxide semiconductor. Within the realm of composite materials, carbon nanotubes and aluminum matrix composite materials (with graphite or SiC) have attracted a lot of interest due to their potential thermal properties. Lastly, the development of an injection moldable polymer is also reviewed.

10.1 Future Trends in Metal Matrix Composite (MMC) Thermal Management Materials

10.1.1 Adjustable CTE and High-TC Thermal Management MMCs

Here we illustrate the unique requirements of heat sinks using packaging of a laterally diffused metal-oxide semiconductor (LDMOS) as an example. There are two major structural categories of RF MOSFETs in use today: double-diffused metal-oxide semiconductor (DMOS) and LDMOS. Both of these MOSFETs are composed of three terminal devices (assuming substrate shorted to source), commonly identified as the source, gate, and drain, where the voltage on the gate controls the current flowing from the drain to the source. The most common circuit configuration for these devices is the common source (CS) configuration. Other configurations are used, but under the CS configuration the drain is connected to the high DC voltage, while the source is grounded.

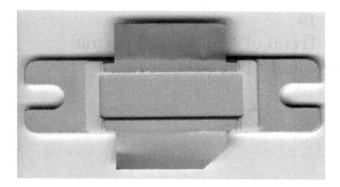

Fig. 10.1 LDMOS heat sink package

In contrast to the LDMOS, the DMOS structure consists of an n-type substrate on which is grown an n-type epitaxial layer that forms the large drain region. The gate region is formed on the surface that overlies the graded p-type body implantation and diffusion area. The source regions are implanted and diffused on either side of the gate to form two separate transistors with a common drain region.

Another fundamental difference between the DMOS and LDMOS is the characteristic of the source connection to the outside world. The source of a DMOS is located at the surface of the die, while the drain is formed by the substrate region, and therefore wire bonds must be used to connect the source to the external circuitry. These wire bonds form a dependent frequency element reducing gain at high frequencies due to negative feedback. In addition, the common surface connection dictates that insulating material (BeO or Al_2O_3) can be used to isolate the drain. This insulating material has a thermal impedance that must be considered for power dissipation. The source of the LDMOS is also on the surface; however, its common connection is formed by diffusing a highly doped p-type region, which acts like an ohmic connection from the source at the surface of the substrate, eliminating the need for parasitic wire bonds. This method also eliminates the ceramic interface (that exists in DMOS) and improves the junction to case thermal resistance and, therefore, relieves some of the associated power dissipation issues.

In the case of DMOS, the insulating material (BeO or Al_2O_3) must be bonded to a heat sink assembly; in the case of LDMOS, the semiconductor substrate must be bonded to a heat sink assembly. And here is one of the great challenges in microelectronics packaging and optoelectronics packaging – the matching of the CTE of dielectric material and substrate materials to those of the heat sink material. Because there are varieties of dielectric material and substrate materials with very different CTEs, it is desirable to have heat sink materials that have CTEs that can be adjustable to minimize the manufacturing cost. Further, because the LDMOS has a strip structure, it is desirable to have a heat sink material that has a high in-plane TC. The Cu/CuMo/Cu (CPC) laminated composite material is well suited for this kind of application (Fig. 10.1).

10.2 Future Trend of Reinforced MMC Thermal Management Materials for Microelectronics and Optoelectronics

10.2.1 Carbon Nanotubes

Carbon nanotubes (CNTs) are among the strongest structures theoretically possible; they have an axial TC approaching 3,000 W/mK and conduct heat exceptionally well along their axes. Tubes are highly resistant to degradation from heat and chemicals. Many tubes show "ballistic transport" of electrons. CNTs, and double-wall nanotubes (DWNTs) in particular, are the world's most efficient producers of field emission electrons. Various research efforts have concentrated on making composites using CNTs as reinforcement. However, the enhanced thermal transmission properties of CNTs have not improved the thermal transmission properties of the composite yet. CNT properties are summarized in Table 10.1 [1, 2].

10.2.2 Aluminium/Diamond Composites

Al/diamond composites exhibit a high TC of 550–650 W/mK, a CTE of 7.5 ppm/°K, very low density of 3.1 g/cm^3, and a melting point of 650°C [3]. They are produced using high-pressure squeeze casting using temperature conditions slightly over the melting point of Al. After sintering to close-to-net shape, parts can be machined using conventional surface grinding or wire/sink EDM. Ni and Au plating have been demonstrated for these composites. The use of synthetic diamond at about $140/lb has reduced the cost structure of these composites. Secondary machining and surface preparation constitute the main cost component for these heat sinks. EDM'ing of through holes and other dimensional features to ± 0.001" has been demonstrated. The surface finish is rough, approximately 400 μin. Precision flatness would require the use of an Al "skin." The feasibility of this technology has been demonstrated. Several parts have been produced in limited quantities, though industrial scale has not been reached. These composites provide an attractive cost/performance ratio indicator.

Other composites using diamond as reinforcement and Cu, Co, Ag, Mg, and Si as matrix have also been produced exhibiting high TC, low density, and low CTE. Properties are shown in Table 10.2 [1]. The use of SiC as matrix for diamond composites is also an attractive formulation.

Al/diamond/graphite composites have also been developed to package silicon carbide (SiC)-based power-switching devices to enable the next generation of high-torque electric motor drives to operate at 200–300°C without the need for active cooling.

The integrated high TC Al/diamond/graphite base plates house high-performance IGBTs. DBC ceramic substrates may be used to carry power-switching devices and can be integrated into the Al/diamond base plate, thereby eliminating the low-reliability

Table 10.1 Carbon nanotube properties [1, 2]

Property	Unit	Value per nanotube type			Alternative fillers		
		Single wall (SWNT)	Double wall (DWMT)	Multi wall (MWNT)	Carbon fiber	Graphite	Steel
Density	g/oc	0.94 max	0.77 max	2.10 max	2.0 max	20 max	8
Diameter	nm	1–5	2.5–6.2	13–50	7,000	–	–
Length	μ	1–30	2–50	10–500	–	–	–
l/d	–	1,000–5000	500–12,000	2000–20,000	–	–	–
Elastic modules	GPa	1,200–1,700	1000–2,000 (estimated)	1,000–3,700	200–700	1,060	200
Tensile strength	GPa	300–1,500	300–2,000	300–600	2–7	–	0.4
Thermal conductivity	W/mK	3,000	1,500–3,000	1,500	20–1,200	1,500	20–70
Resistivity	μΩ.cm	0.03	0.03–0.1	0.1	100	40	10
Strain to failure	%	20–40	20–40	20–40	2	–	2.5

Table 10.2 High-performance diamond composites [1]

Matrix	Reinforcement	Inplane TC TC (W/mK)	Through TC (W/mK)	CTE ppm/°K	Density g/cm³
Al	Graphite flake	400–600	80–110	4.5–5.0	2.3
Al	Diamond particle	550–600	550–600	7.0–7.5	3.1
Al	Diamond and SiC part	575	575	5.5	–
Cu	Diamond particle	600–1,200	600–1,200	5.8	5.9

solder interface used in commercial power modules. These substrates can be directly integrated with a passively cooled heat sink and enable at least 5× reduction of thermal resistance from heat sink to the device junction.

A low thermal resistance packaging/die attachment technique based on transient liquid phase (TLP) bonding and capable of withstanding high temperatures is allowed by the use of SiC devices and achieves significant reduction of the temperature gradient between the junction and the ambient. The TLP approach achieves 10× reduction in thermal resistance compared to soldered attachments and improved reliability due to increased resistance to the formation of voids under thermal and power cycling conditions. The final package solution allows the manufacture of power modules within custom enclosures without requiring active cooling. These packages are intended for three-phase motor inverters. Robust power modules consisting of silicon IGBT power-switching devices that allow maximum junction temperature rise to 175 °C, and SiC Schottky diodes capable of operating at high junction temperatures up to 300 °C are used. The SiC Schottky diode section of the power module is cooled via natural convection only while the silicon IGBT can be passively cooled or air-cooled depending on the application. The passively cooled hybrid power module allows substantially larger output power delivered to the motor, as compared to the current state-of-the-art power modules. This is due to reduced thermal resistance and reduced losses in the Si IGBT and SiC diodes due to elimination of reverse recovery losses. Several hybrid power modules can be paralleled to achieve higher output power.

This advanced packaging technology will allow for hybrid power modules consisting of SiC Schottky diodes and Si IGBTs. The power module layout will be designed such that the SiC power switching devices can be drop-in replacements (with minor modifications in the gate drive circuitry) for the Si IGBTs. These are becoming more available.

SiC-based power-switching devices such as MOSFETs and IGBTs with high current ratings are currently in the research-and-development stage and are not widely commercially available. The technology to fabricate high current SiC Schottky diodes, however, is quite mature and diodes with current ratings greater than 20A are commercially available. SiC Schottky diodes do not exhibit the reverse recovery behavior observed in Si PiN diodes. Reverse recovery loss contributes to 60–70 % of the power loss in a switching device during turn-on. Hence, replacing just the Si PiN diodes by SiC Schottky diodes significantly reduces the power loss in the switching devices, resulting in a substantial improvement in efficiency.

Reduced power loss in the switching devices also results in reduced junction temperature rise, simplified cooling, higher power density, and greater reliability. The switching loss in the IGBTs can be reduced by 2× by elimination of diode reverse recovery. These inverter modules will be used in a modular design to produce 50- to 100-kW motor drives.

10.3   Injection Molded Polymer Materials for Microelectronics and Optoelectronics

Low cost microinjection molding technology is under development, and Fuh et al. have proposed an integrated manufacturing methodology to use micromolding techniques and electroplating processes to fabricate 3D structures combining antennae, splitters, waveguides, and other passive mechanical components in a batch process fashion. The dramatically different architecture as compared to the traditional system is the batch process to reduce the manufacturing cost. To make a versatile front-end manufacturing process, we propose to build plastic manufacturing modules, including plastic molding, embossing, mechanical polishing, assembly, and bonding processes as shown in Fig. 10.2. This process eliminates the current manual-assembly requirement of bulky mechanical reflectors and phase-shifters/ amplifiers in single-mode millimeter-wave waveguides to reduce cost, size, and operation power. The key advantage of this process as compared with other processes, such as stereolithography or EFAB [3–5] is that this approach is better suited for millimeter-wave systems, such as antennae, waveguides, and coaxial transmission

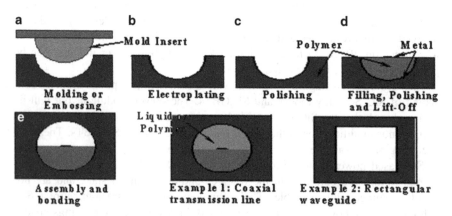

Fig. 10.2 Cross-sectional views of various process modules for the 3D, plastic-metal manufacturing process for integrated millimeter-wave systems, including (**a**) molding or embossing of plastic substrates, (**b**) electroplating of metals, (**c**) mechanical surface polishing, (**d**) new polymer filling, polishing, metal lift-off, or patterning, and (**e**) assembly and bonding for integration. Two millimeter-wave components are shown, including a 3D coaxial transmission line and a rectangular waveguide.

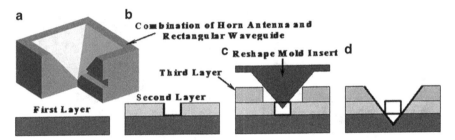

Fig. 10.3 Fabrication example of proposed process modules. The horn-shaped antenna is to be fabricated as part of the integrated system. The second layer substrate is molded to have the desired rectangular waveguide shape and bonded with the first layer (**a**). An electroplating process is conducted and followed by a mechanical polishing process to clean the metal on the surface as shown in (**b**). The third layer is first molded to have the desired shape as illustrated and the back of the layer is deposited and patterned with a layer of metal to provide the top surface of the waveguide structure. It is then bonded with the system and a reshape mold is applied to define the horn antenna in (**c**). The system is completed by final electroplating and polishing process in (**d**)

lines. First, these components are relatively big 3D structures (up to the millimeter range), and it would take a rather long processing time for the competitive thin-layer processes to construct these structures. Second, multiple alignment steps are expected by using thin-layer-based processes such that the sidewalls of waveguides are expected to be rough. A "three-thick-layer" architecture for the 3D antenna array can be fabricated based on the process modules as illustrated in Fig. 10.2. The proposed thick-layer manufacturing process is versatile and can be applied to devices beyond this work. For example, the two fabrication examples in Fig. 10.2 of circular coaxial and rectangular waveguides use "two-thick-layer" structures. Another example of a three-thick-layer architecture for the 3D antenna array can be fabricated based on the same process modules as illustrated in Fig. 10.3, where the horn-shaped antenna is to be fabricated as part of the integrated system.

Figure 10.4 illustrates a further embodiment of a schematic 4×4 millimeter-wave radar system using the rectangular waveguide and waveguide-based components in the feeding networks. For W-band systems, the dimension of a waveguide is 2.54×1.27 mm^2, and the fabrication process is typically done by precision machining on a metallic block followed by brazing a metallic cover to enclose the waveguide. In the new polymeric molding process as illustrated in Fig. 10.4, an aluminum mold insert is fabricated by mechanical machining processes and used to shape a polymeric substrate in the molding process to form an open waveguide at a temperature above the glass transition point of the chosen polymer. The polymeric replica is then detached and a metallic seed layer, such as 200 Å/6,000 Å Cr/Pt, is sputtered for the electroplating step. A second substrate is sputtered with the same seed layer and is clamped on top of the first substrate to cover the open waveguide structure. The clamped combination is then immersed into a gold electroplating bath to deposit an 8-μm-thick gold layer. The selective electroplating process also metallically seals the waveguides.

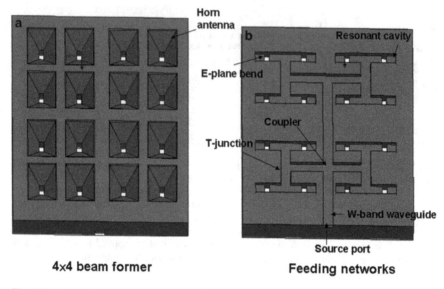

Fig. 10.4 A 4×4 beam-former horn array illustrating the architecture of the 3D millimeter-wave radar system. (**a**) Top view showing main system components including horn array antennas. (**b**) Sectional view showing waveguides, couplers, T-junctions, and resonant cavities

The fabrication process of polymeric MEMS millimeter-wave systems can be classified into low- and high-aspect-ratio operations according to the structural geometries of the various components. For low-aspect-ratio components such as waveguides and waveguide-based iris filters, the process is illustrated in Fig. 10.5a–c, where micromolding, selective electroplating, and assembly processes are routinely used. High-aspect-ratio structures such as waveguide-fed horn antennae use a self-aligned 3D hot embossing molding process, followed by two-step selective sputtering (front and back side of polymer substrate) and in-channel electroplating (Fig. 10.5d–f). There are two key technology challenges that will require careful investigation. First is the plastic-to-plastic bonding process in which the bonding strength and integrity is of great importance. This critical metal bonding process is accomplished by the application of selective electroplating and bonding process developed previously [5, 7].

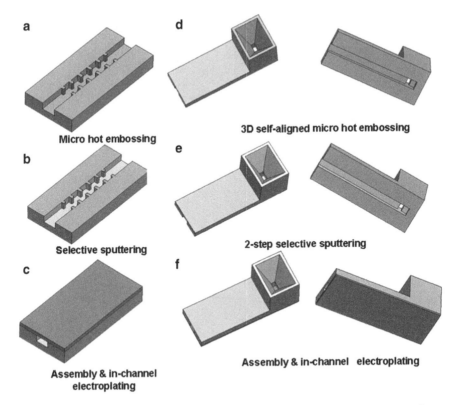

a Micro hot embossing

b Selective sputtering

c Assembly & in-channel electroplating

d 3D self-aligned micro hot embossing

e 2-step selective sputtering

f Assembly & in-channel electroplating

Fig. 10.5 Fabrication process of typical polymeric MEMS millimeter-wave components. Low-aspect-ratio components such as waveguide and waveguide-based iris filters and polymeric wave-guides are illustrated in (**a**)–(**c**). In process (**a**), a polymer substrate is molded to create the waveguide shape. (**b**) Selective electroplating to deposit metallic seed layer. (**c**) A second polymer substrate is assembled by means of selective electroplating and bonding process. High-aspect-ratio structures such as waveguide-fed horn antennae are fabricated in steps (**d**)–(**f**). (**d**) Front-side and back-side views of a polymeric substrate fabricated by means of self-aligned 3D hot embossing process using both front-side and back-side mold inserts. (**e**) Selective seed layer deposition in the front and back side. (**f**) A second polymer substrate is assembled by means of selective electroplating and bonding process

References

1. Sepulveda JL (2010) Polymeric high thermal dissipation ceramics and composite materials for microelectronic packaging. In: Kuang K, Kim F, Cahill S (eds) RF and microwave microelectronics packaging. Springer, New York
2. Loutfy RO, MER Corporation, Tucson AZ (2011) www.mercorp.com
3. Fuh Y-K, Sammoura F, Jiang Y, Lin L (2010) Polymeric microelectromechanical millimeter wave systems. In: Kuang K, Kim F, Cahill S (eds) RF and microwave microelectronics packaging. Springer, New York
4. http://www.isi.edu/efab
5. Pan LW, Lin L (2001) Batch transfer of LIGA microstructures by selective electroplating and bonding. IEEE/ASME J Microelectromech Syst 10:25–32

Index

G. Jiang et al., *Advanced Thermal Management Materials*,
DOI 10.1007/978-1-4614-1963-1, © Springer Science+Business Media New York 2013